Progress in Optical Science and Photonics

Volume 6

The purpose of the series Progress in Optical Science and Photonics is to provide a forum to disseminate the latest research findings in various areas of Optics and its applications. The intended audience are physicists, electrical and electronic engineers, applied mathematicians, biomedical engineers, and advanced graduate students.

More information about this series at http://www.springer.com/series/10091

Sreekanth K. V. · Mohamed ElKabbash ·
Vincenzo Caligiuri · Ranjan Singh ·
Antonio De Luca · Giuseppe Strangi

New Directions in Thin Film Nanophotonics

Springer

Sreekanth K. V.
School of Physical and Mathematical Sciences
Nanyang Technological University
Singapore, Singapore

Vincenzo Caligiuri
Nanochemistry Department
Istituto Italiano di Tecnologia
Genoa, Italy

Antonio De Luca
Department of Physics
Universita Della Calabria
Rende, Cosenza, Italy

Mohamed ElKabbash
Institute of Optics
University of Rochester
Rochester, NY, USA

Ranjan Singh
School of Physical and Mathematical Sciences
Nanyang Technological University
Singapore, Singapore

Giuseppe Strangi
Department of Physics
Case Western Reserve University
Cleveland, OH, USA

ISSN 2363-5096 ISSN 2363-510X (electronic)
Progress in Optical Science and Photonics
ISBN 978-981-13-8890-3 ISBN 978-981-13-8891-0 (eBook)
https://doi.org/10.1007/978-981-13-8891-0

This Springer imprint is published by the registered company Springer Nature Singapore Pte Ltd.
The registered company address is: 152 Beach Road, #21-01/04 Gateway East, Singapore 189721, Singapore

Preface

Recent advances in nanophotonics demonstrated remarkable control over the electromagnetic field by tailoring the optical properties of materials at the subwavelength scale which resulted in the emergence of the fields of metamaterials and metasurfaces. However, the control and manipulation of electromagnetic waves in most cases require intensive structuring at the nanometer scale of the materials, which can only be achieved with advanced and high-fidelity nanofabrication process. Stacked ultra-thin films of dielectrics, semiconductors, and metals were introduced as an alternative platform that performs unique or similar functionalities. The new era of thin-film photonics departs from the traditional domain of thin films as mere interference optical coatings with many potential applications including perfect and selective light absorption, structural coloring, biosensing, gas sensing, enhanced spontaneous emission and as superlenses. In addition, the discovery of the interesting optical properties of chalcogenide-based phase change materials has opened the door for photonic data storage again because the optical constants of these materials can be rapidly activated thermally, optically, or electronically. The versatile and rapid modulation of such materials makes them suitable for a wide range of reconfigurable photonic device applications.

This book highlights the recent advances in thin-film nanophotonics, particularly, and discusses metal/dielectric resonant cavity-based thin-film structures and metal/dielectric multilayered hyperbolic metamaterials. The potential applications for these thin-film structures in ultra-sensitive biosensing, gas sensing, perfect light absorption, engineering of spontaneous emission enhancement, structural color filters, and reconfigurable photonic devices are presented.

Part I of the book discusses various resonant cavity-based thin-film structures. This includes the demonstration of critical light coupling in thin- and ultra-thin-film structures, structural coloring via thin films, hydrogen gas sensing, experimental realization of extreme phase singularity at the point-of-darkness using asymmetric Fabry–Perot cavities, and the demonstration of tunable perfect absorption in the visible spectral region using phase change chalcogenide-based nanophotonic cavities. In addition, we discuss the realization of enhanced Goos–Hänchen shift using resonant cavity-based thin-film structures. Finally, we show how to use interfering surface plasmons—excited in thin metal films—to create subwavelength periodic nanostructures.

Part II of the book investigates the thin-film nanophotonics as a building block of metamaterials and metasurfaces. In particular, we present the design, fabrication, and applications of various multilayered hyperbolic metamaterials (HMMs), which exhibits hyperbolic dispersion in different spectral bands. The first two chapters of Part II discusses the realization of dielectric and resonant gain singularities in HMMs. In addition, we discuss the applications of dielectric and resonant gain singularities in HMMs for supercollimation of light and superlensing. Part II also discusses the control of topological transition using gain-embedded HMMs. We further demonstrate the excitation of HMM-guided modes using grating and prism coupling techniques and its application for ultra-sensitive biosensing, spectrum selective perfect absorption, and spontaneous emission enhancement. At the end of Part II, we propose a strategy to realize negative refraction in terahertz frequencies using graphene and topological insulator-based HMMs. It is shown that the proposed thin-film stacks based on resonant cavities and HMMs are a simple and scalable platform for various potential applications including ultra-sensitive biosensing of small biomolecules, hydrogen gas sensing at low sample volumes, single photon sources, and active photonic devices.

This book is proposed to be used as a reference book for graduate students and researchers working in the photonics field including thin-film optics, biosensing, phase change materials, nanophotonics, plasmonics, and metamaterials. The central aim of this book is to highlight recent advances in active thin-film photonics and to create a bridge toward novel and advanced applications as metamaterial machines, active superlenses, and quantum realm platform, among others.

Singapore — Sreekanth K. V.
Rochester, USA — Mohamed ElKabbash
Genoa, Italy — Vincenzo Caligiuri
Singapore — Ranjan Singh
Rende, Italy — Antonio De Luca
Cleveland, USA — Giuseppe Strangi

Acknowledgements We would like to thank Prof. Ken-Tye Yong and Prof. Murukeshan Vadakke Matham, Nanyang Technological University, Singapore. We acknowledge the support of Prof. Michael Hinczewski and Prof. Umut A. Gurkan at Case Western Reserve University, USA. We would like to extend our thanks to Prof. Robert Simpson at Singapore University of Technology and Design, Singapore, and Dr. Sreejith Sivaramapanicker at National University of Singapore.

Contents

Part I
Development and Applications of Metal/Dielectric Resonant Cavity-Based Thin Film Structures

Miniaturized optical devices with strong responses are required to properly compete with electronic storage. Optical response of the materials can be amplified when the light is critically coupled to a resonator. In particular, light is critically coupled when reflection and absorption rates of a resonator are equal. The perfect absorption of light occurs at the critical coupling, which is due to the existence of nonzero losses in optical resonators. In Part I, we discuss the experimental demonstration of critical light coupling in different metal/dielectric resonant cavity-based systems and their applications in structural coloring and hydrogen gas sensing. More importantly, we demonstrate the realization of singular phase at the Brewster angle and its application for ultra-sensitive biosensing. We present achieving multiband tunable perfect absorption and enhanced Goos–Hänchen shift in the visible spectral region using phase change material-based cavities. In addition, Part I discusses plasmonic lithography concept based on surface plasmon interference generated by thin metal films.

Chapter 1
Perfect Light Absorption in Thin and Ultra-Thin Films and Its Applications

Perfect light absorption (PLA) has a wide range of applications from solar-thermal based applications to radiative cooling. However, most of the proposed platforms require intense lithography which makes them of minor practical relevance. In this chapter, we will first discuss the different approaches to realize perfect light absorption. In Sect. 1.3, we will introduce interference based thin-film light absorbers using two, three, and multi-layer films. In Sect. 1.4, we provide a blueprint for designing light absorption in thin films in terms of wavelength range, bandwidth, angular range, and the spatial profile of light absorption. Section 1.5 discusses the iridescence properties of thin film based light absorbers. Section 1.6 discusses thermally induced perfect light absorption in metals. Finally, we discuss, in details, demonstrated applications (Sect. 1.7) of thin film light absorbers in structural coloring and hydrogen gas sensing.

1.1 Introduction

Suppressing reflection, using anti-reflection (AR) coatings, has been the main objective in early works on thin-film optics. For most conventional applications, AR coatings aim is to maximize transmission particularly when there are multiple optical elements, e.g., compound and zoom lenses, that lead to low net transmission [1]. Accordingly, the developed coatings consisted of a single thin-film or a stack of thin-films with a quarter-wavelength optical thickness and no absorptance (A) to achieve zero-reflectance and near unity transmittance at a given wavelength range. In fact, optical losses in the form of light absorption are considered a major obstacle that hinders the progress of many applications. For example, in the field of plasmonics and metamaterials, the problem of strong optical losses is the major obstacle in applications and devices that require high efficiency, e.g. perfect lenses, clocking devices, and plasmonic transistors and interconnects [2].

However, a lot can be gained from optical losses where efficient reflection suppression over a given wavelength range, with minimal footprint, is important

Sreekanth K. V. et al., *New Directions in Thin Film Nanophotonics*, Progress in Optical Science and Photonics 6, https://doi.org/10.1007/978-981-13-8891-0_1

while transmission is of no value, when light-to-heat conversion is desired, or in order to increase the emissivity (thermal radiation) of a surface. Accordingly, suppressing reflection and transmission to realize strong (A > 0.9) to perfect (A > 0.99) light absorption has been of great importance [3]. Various applications that require engineering light absorption were developed [4] including structural coloring [3, 5–9], heat-assisted magnetic recording (HAMR) [10], cancer therapy [11], photo-acoustic imaging [12], solar thermophotovoltaics [13], solar thermal power generation [14], solar thermoelectric generation [15], solar steam generation [16], solar based water sanitation and desalination [17], radiative cooling [18], photodetectors [19], and optical switching [20].

1.2 Approaches to Realize Perfect Light Absorbers

The "recipe" to realize perfect light absorption has two requirements: (i) to minimize reflection by admitting the electromagnetic field in a medium, (ii) such medium must have non-zero losses such that the entire field energy is dissipated within the medium. This begs the question: is it possible to realize perfect or strong light absorption using a smooth, unpatterned substrate (Fig. 1.1a)? Figure 1.1b shows the calculated absorptance of normally incident light with wavelength $\lambda = 700$ nm from air on an infinitely thick substrate for different values of the real (n_s) and imaginary (k_s) parts of the complex refractive index $\widetilde{n_s} = n_s + ik_s$ [21]. For values of approximately $n_s = [0.5, 2]$ and $k_s = [0.01, 1]$, strong light absorption is possible. These values correspond to a refractive index that is very close to that of air, and small, but nonzero, extinction coefficient k_s. A material with these optical constants, thus, has low impedance mismatch with the incident medium, air, and the electromagnetic field is admitted inside the material. Furthermore, such material would have nonzero optical losses and, eventually, would absorb all the incident electromagnetic energy. The optical constants of several materials are shown in Fig. 1.1b and the corresponding absorptance values closely agree with the absorptance measured at 700 nm for these materials. As can be seen, none of these materials lie in the strong light absorption region for 700 nm. The lack of strong light absorption in bulk materials is due to the strong impedance mismatch between the incidence medium and the absorbing medium.

Semiconductors that have a bandgap at a given wavelength range, e.g., Ge and Si in the visible, have strong light absorption due to inter-band transitions, i.e., $k_s \neq 0$. However, the bandgap is associated with strong dipolar response from the constituent atoms, thus, they have large n_s., i.e., they have strong reflection as they are not impedance matched with air. On the other hand, metals have small n_s but extremely high losses due to the existence of free electrons. Accordingly, impedance is severely mismatched and most of the light is reflected.

On the nanoscale, however, interesting physics emerges. Metallic nanostructures exhibit extremely strong light absorption due to the excitation of resonant modes called localized surface plasmons. At resonance, electromagnetic field is strongly

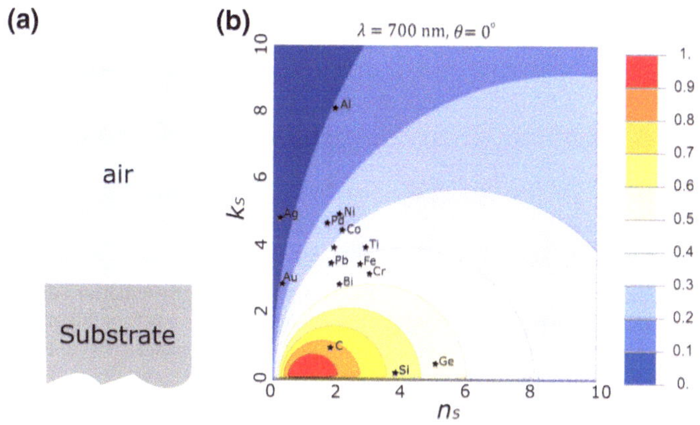

Fig. 1.1 **a** A schematic of homogeneous infinite substrate with a complex refractive index $\widetilde{n_s} = n_s + ik_s$. **b** The calculated absorptance of normally incident light with $\lambda = 700$ nm for the infinite substrate depicted in (**a**), the index $\widetilde{n_s}$ of different materials is marked. Known materials do not exhibit strong light absorption. Reproduced with permission from Ref. [3], John Wiley and Sons

localized within the nanoparticle as energy is transferred from the electromagnetic field to the kinetic motion of free electrons (Fig. 1.2a) [2]. This is the so called lightening rod effect [22]. Because free electrons in nanoparticles suffer from strong electron-electron and electron-phonon scattering, as well as, from Landau damping [2], metallic nanostructures enable field localization in a lossy medium and can realize strong light absorption. Indeed, various metallic nanostructures have been used to realize perfect light absorption for photothermal and solar-thermal applications [23]. Thc plasmon resonance not only depends on the metal's optical properties, the nanostructure size, and the surrounding environment, but also depends on the interaction with other metallic nanostructures. The interaction between the near fields of nanoparticles results in so called plasmon hybridization which can broaden the plasmon resonance with some tunability [23]. On the other hand, high-index semiconductor nanowires, e.g., Ge nanowires, have shown strong light absorption that can be tuned by tuning the nanowire dimensions due to the excitation of leaky-mode resonances [24]. These modes can be thought of as light trapped in a high-index nanowire via total internal reflection. However, these modes are difficult to control as their near-field does not couple strongly which exclude the realization of broadband light absorption [25].

Metamaterials are engineered materials that enjoy optical properties and functionalities beyond what natural materials can provide. Usually metamaterials are composed of different materials and/or structures that composes artificial atoms which interact with light resulting in an emergent property due to the interplay of all the component materials and/or structures. An important class of metamaterials is left-handed metamaterials where the meta-atoms support electric and magnetic dipoles in any wavelength range, even if the constituting materials do not support a magnetic dipole. These materials can have, simultaneously, a negative permittivity and

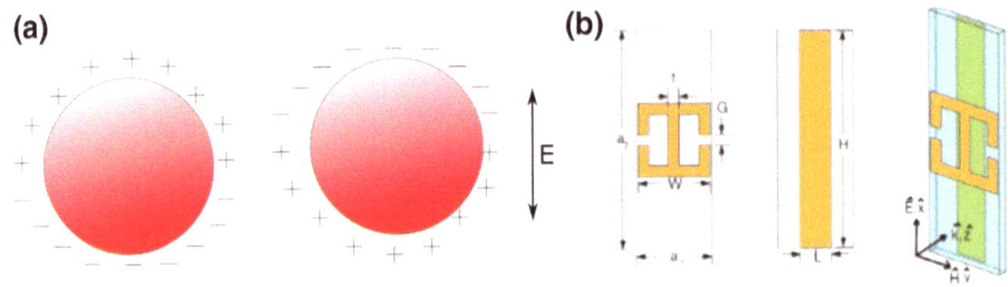

Fig. 1.2 **a** Schematic of a metallic nanoparticle that supports charge oscillations due to irradiation with an electric field at the plasmon resonance frequency. **b** Shows a schematic of an electric ring resonator and a cut wire that supports electric and magnetic resonances. (**b**) is reproduced with permission from Ref. [26], American Physical Society

negative permeability, i.e., a negative refractive index. So far, negative index materials, however, failed to live up to the expected applications, e.g., perfect lenses and invisibility cloaks, due to optical losses. A famous example of negative index meta-atoms is the split ring resonator with a cut wire which is, in its essence, an LC circuit (see Fig. 1.2b) [26]. These meta-atoms enable tuning the electric resonance and, consequently, the effective electric permittivity ε. Furthermore, the oscillating currents in the resonator couples to the magnetic field at the wavelength of choice and allows tuning the magnetic permeability μ. For a magneto-dielectric medium, the reflectance is given by $R = |Z - Z_0/Z + Z_0|^2 = |\mu - n/\mu + n|^2$, where the impedance $Z = \sqrt{\mu/\varepsilon}$, the free space impedance $Z_0 = \sqrt{\mu_0/\varepsilon_0}$, μ_0 and ε_0 are the free space permeability and permittivity, respectively, and $n = \sqrt{\mu\varepsilon}$. Clearly, when $Z = Z_0$, $R = 0$. Accordingly, by tuning μ and ε a left-handed metamaterial can realize zero reflection. Furthermore, because the metamaterial has finite losses, light can be fully absorbed inside the metamaterial [4, 26].

However, metamaterials require intense nanofabrication at visible and NIR wavelength ranges since the constituting meta-atoms must be significantly smaller than the wavelength range of interest. This leads to our topic of interest, thin films for perfect light absorption. The deposition of thin films is a matured technology that is relatively simple and highly scalable. Thin film light absorbers are antireflection coatings with lossy materials, e.g., metals and semiconductors. In the next section, we will discuss the different strategies to create perfect light absorption using thin films.

1.3 Lithography-Free Perfect Light Absorption in Critically Coupled, Interference Based, Thin and Ultrathin Films

To realize perfect light absorption, light should be critically coupled to the absorber [27, 28]. Critical light coupling takes place when absorption equals the sum of the reflection, transmission, and scattering. For relatively thick, opaque materials, e.g., metals with thickness in the order of 100 nm, the transmission is negligible. Furthermore, for smooth thin films, scattering is also negligible. Accordingly, to realize critical light coupling in thin films, it is sufficient to balance absorption and reflection. Suppressing reflection in thin films is realized via thin film interference.

Critical light coupling in thin films takes place due to amplitude splitting destructive interference [3] such that light is entirely trapped inside the resonator and is dissipated due to the existence of losses. Accordingly, two conditions must be simultaneously satisfied to realize perfect absorption; the interfering waves must be out of phase (phase condition), and the out-of-phase waves must be of equal amplitude (amplitude condition) (see for more details Ref. [19]). Meeting these conditions means that light is trapped inside a lossy medium and, eventually, the field energy will be dissipated entirely, i.e., perfect light absorption is realized. These conditions can be met using different configurations; an asymmetric metal-dielectric-metal Fabry-Perot cavity, a lossy dielectric on a high reflectance metal, and a lossless dielectric on a low reflectance substrate.

Satisfying the amplitude condition is straightforward in asymmetric Fabry–Perot metal–dielectric–metal (MDM) cavities as the top and bottom metal layers are highly reflective and the amplitude can be split among the two metal layers. The phase condition is satisfied by ensuring that the dielectric layer optical thickness $t_{Opt} = n_d t = \lambda/2$, i.e., cavity modes are where n_d is the dielectric refractive index and t is the physical thickness of the dielectric. MDM cavities, thus, offer a simple route to obtain perfect light absorption. Note that if the bottom metal layer is not opaque, the MDM cavity acts as a color filter where the resonance corresponds to low reflectance and high transmittance [29]. MDM cavities have variety of applications including structural coloring and gas sensing. However, the existence of a top-metal layer has several drawbacks, e.g., it can easily suffer from scratching and oxidation which will deteriorate the formed colors rapidly. Furthermore, there are many applications that require concentrating light absorption in a single metal or dielectric layer, e.g., photovoltaics, solar-thermal applications, and photo-detection. However, the absorption in an MDM cavity is distributed between the two metal layers. Finally, for longer wavelengths, using a thin film that is approximately half the wavelength can be costly and difficult to deposit.

In general, it is more difficult to satisfy the amplitude condition using a dielectric coating on a metallic substrate given that dielectrics do not strongly reflect light. Using a lossy dielectric on a high reflection metal, however, can satisfy both the phase and amplitude conditions, while maximizing optical absorption inside the semiconductor layer. The lossy dielectric dampens the magnitude of the partially reflected waves from the highly reflective metallic substrate, thus, the reflection amplitude from both the dielectric and the metal layer can be equal (for more information, see [19]). The phase condition for dielectric thicknesses $t_{Opt} < \lambda/4$. This is because the required phase difference for destructive interference is accumulated not only due to the propagation inside the dielectric but also due to the non-trivial phase shift $\Delta\emptyset$ upon reflection from the air-lossy dielectric, and lossy dielectric-metal interfaces. This is because $\Delta\emptyset \neq 0, \pi$ when light reflects off a lossy dielectric and when the metal is not behaving as a perfect electric conductor (which is usually the case for low-reflection metals beyond mid IR range and high-reflection metals beyond the NIR range). Because the required t_{Opt} is significantly less than that in an MDM cavity, the ultrathin lossy dielectric-reflective substrate approach is particularly advantageous for IR applications.

However, requiring that the dielectric is lossy, and the metallic substrate is highly reflective but not a perfect electric conductor to realize perfect light absorption using a thin dielectric film limits the wavelength range and materials that can be used to realize strong to perfect light absorption. Most dielectrics are lossless, and semiconductors absorption is, mainly, limited to wavelength ranges where interband transitions occur. Furthermore, high reflection metals, e.g. Ag, and Au, become perfect electric conductors in the NIR.

Using a lossless dielectric would be ideal, however satisfying the amplitude condition becomes problematic as the field is not attenuated inside the dielectric and most dielectrics have low n_d particularly in the visible and NIR ranges. Consequently, the amplitude condition requires using a low reflectance substrate, i.e., lower the n_d (and reflection from the dielectric layer) necessitates a substrate with lower reflection and vice versa (Fig. 1.3a). Furthermore, the non-trivial $\Delta\emptyset$ is acquired from reflection off the dielectric-low reflectance substrate. Accordingly, $t_{Opt} < \lambda/4$ is still possible in this configuration. Figure 1.3b shows the modified absorptance as a function of the substrate optical constants (n_s and k) when adding a lossless dielectric coating with $t = 40\,\text{nm}$ and $n_d = 2.5$. The absorptance did not increase significantly for substrates with high reflectance, for example, Al, Au, and Ag. On the other hand, the absorptance significantly increased for many substrates with higher absorptance, for example, Ni, Pd, Ti, Fe, Pb, Cr, Bi, and Co. Accordingly, strong to perfect light absorption is possible for lossless dielectrics with $t_{Opt} < \lambda/4$, (here $t \approx \lambda/7n_d$) on a variety of substrates. Note that the absorptance did not increase significantly for substrates with significant absorptance (A > 0.5), for example, C, Si, and Ge. Accordingly, having an absorptive substrate per se is insufficient to realize perfect light absorption without careful investigation of the dielectric coating parameters contrary to what is indicated in Ref. [19].

Finally, an interesting scheme of light trapping that is capable of increasing the absorption cross-section of ultrathin absorbing films is to use a lossless dielectric on a high reflectance metal while adding an ultrathin absorbing layer. Without the absorbing layer, this optical coating can only modify the reflected light phase with negligible intensity modification, i.e., act as a Gires-Tournois etalon [7]. However, adding the ultrathin absorbing film allows the trapped light to be dissipated with high power dissipation density values inside the ultrathin absorbing film. This technique was introduced in [28] where the authors used a distributed Bragg reflecting mirror and a lossless dielectric to realize amplitude-splitting destructive interference, and a 5 nm thick polymer doped with strongly absorbing J-aggregate dye molecules to realize PLA. Furthermore, this mechanism is why graphene layers, with $A\sim 2.3\%$, appear clearly when placed on a Si substrate. This is because the Si substrate has a natural oxide on top and both realize destructive interference which traps light that gets dissipated significantly inside the graphene layer. The same technique was

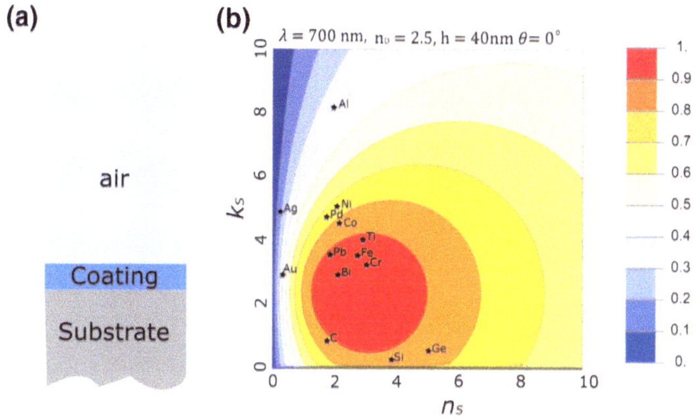

Fig. 1.3 **a** A schematic of a homogeneous ultrathin lossless dielectric coating on an infinite substrate. **b** The calculated absorptance of normally incident light on the lossless dielectric-substrate structure for different values of n_s and k_s. The range where strong to perfect light absorption takes place increased, compared to that shown in Fig. 1.1b, and includes several materials. Reproduced with permission from Ref. [3], John Wiley and Sons

further used for ultrathin semiconductors in [30] and to maximize the fluorescence from two-dimensional direct bandgap materials, e.g., MoS_2 [31]. By precedence, we consider the work presented in [28] to be the pioneering work in critically coupled ultra-thin film perfect light absorbers.

1.4 Designer Perfect Light Absorption in Thin Film Absorbers

One possible drawback of thin-film light absorbers is the difficulty to control the absorption properties compared to, e.g., metamaterials. However, we showed that it is possible to control various aspects of light absorption in thin films including the wavelength range, bandwidth, light absorption angle, spatial profile of optical losses, and iridescence. The designer properties of light absorption were highlighted in various works by us [3, 27] and others [29, 32, 33].

1.4.1 Designer Wavelength Range

Using an MDM cavity allows tuning the wavelength range where light absorption takes place by simply changing the dielectric thickness such that cavity modes when $t_{Opt} = m\lambda/2$ is satisfied, where m is a non-zero integer. We note here that the top metal layer needs to be adjusted as metals tend to be highly reflective at longer

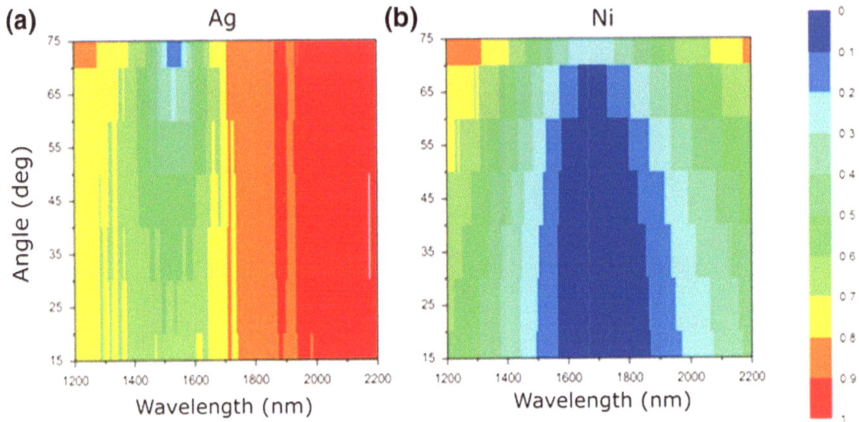

Fig. 1.4 SWIR optical absorber: The absorptivity of an 80 nm thick Ge layer is deposited on **a** Ag and **b** Ni substrates. SLA is only possible using a Ni substrate since Ge has low losses in the SWIR range. Reproduced with permission from Ref. [3], John Wiley and Sons

wavelengths and, accordingly, thinner metallic top layers are required for longer wavelengths.

As we mentioned earlier, it is difficult to find a suitable lossy dielectric at all wavelengths ranges particularly when we are dealing with applications that require non-toxic and low-cost elements since many semiconductors are very expensive and some are toxic. Furthermore, at NIR and longer wavelengths, highly reflective conventional metals, e.g. Au, Ag, Al, cannot be used as back-reflectors, because they are too close in behavior to perfect electric conductors and the dielectric-metal phase shift is ~π. There are several materials, however, which have complex refractive indices in certain regions of the infrared range that are similar to those of conventional metals at higher frequencies, including polar dielectrics, highly-doped semiconductors, and conducting oxides ([19] and references therein). This was demonstrated in [32] using a polar dielectric, sapphire (Al_2O_3), and polycrystalline vanadium dioxide (VO_2) as the lossy layer. This work is particularly interesting as it demonstrated tunable light absorption as VO_2 underwent a gradual insulator-metal phase transition which was thermally triggered.

On the other hand, using a lossless dielectric on a low reflectance substrate allows for a broader wavelength ranges where perfect light absorption can be realized [3]. Figure 1.4a and b show the absorptance for incident angles 15°–75° of a 75 nm Ge layer on Ag and Ni substrates in the short-wave infrared (SWIR) range. Since Ge is not highly absorbing above the SWIR region, strong absorption is realized only for the Ge–Ni structure. In addition, the Ge layer thickness $t \approx \lambda/6n_{Ge}$, for n_{Ge}, ≈4 at 1750 nm. The ultrathin Ge thickness is due to the nontrivial ΔØ upon reflection from Ge/Ni interface.

1.4.2 Designer Absorption Bandwidth

The absorption bandwidth broadness determines the type of application a given light absorber is suitable for. For example, photodetectors would require strong light absorption within a narrow bandwidth, while solar-thermal applications require relatively broadband light absorption that covers, at least, the visible and NIR spectrum. Selective emitters and radiative cooling require the control over the absorption bandwidth so that the emission matches, e.g., the atmospheric radiation window.

Using ultrathin light absorbers (lossy dielectric-high reflectance substrate and lossless dielectric-low reflectance metal systems) offer relatively broadband light absorption for s or p light polarization (see Fig. 1.5a and b) with limited control over the bandwidth. The bandwidth of an MDM Fabry-Perot cavity, however, depends on how well defined a mode is inside the cavity and the optical losses associated with the materials constituting the cavity. Accordingly, by modifying the top metal thickness, it is possible to tune the absorption bandwidth to some extent [29] from narrowband absorption to relatively broadband absorption. Figure 1.5c shows an Ag–SiO2–Ag MDM cavity with $t = 160$ nm and the bottom Ag thickness is $\sim$100 nm, for three different top Ag layer thickness of 15, 30, and 45 nm. Evidently, the bandwidth of thicker top metal films is significantly narrow and reaches ~8 nm. Furthermore, using low reflectance metals at least as one of the MDM mirrors yields broadband light absorption. Figure 1.5d shows a schematic of a broadband MDM light absorber using a Cr–SiO_2–Cr MDM cavity due to the strong losses in Cr [33]. The absorption results are shown in Fig. 1.5e which shows broadband light absorption over the entire visible range.

Creating selective solar absorber selective absorbers are important component in solar-thermal applications. This is because a solar absorber functions best by absorbing wavelengths within the solar spectrum while having minimum absorption/emission at the object's blackbody radiation wavelength range at a given temperature. For this, a broadband absorber that covers the entire visible and NIR spectrum ~(300–2500 nm) is necessary. Multilayer thin films form a cascaded cavity which allows for broadband light absorption due to hybridization of cavity modes. The multilayer thin film can be periodic or aperiodic. Periodic multilayer thin films was demonstrated in [34] who optimized light absorption from periodic alternating layers of tungsten (W) and alumina (Al_2O_3). An aperiodic multilayer absorber structure consisting of SiO_2 (105 nm)/Ti (15 nm)/SiO_2 (95 nm)/Al (100 nm) demonstrated in [35] which exhibited average absorptivity over 0.95 in the visible at normal incidence and over 0.90 up to an incident angle of 60$\pm$. Aperiodic multilayer films are thus simpler to fabricate. Furthermore, using an antireflective coating on the multilayer stack or by decreasing grading (decreasing) the refractive index from the substrate to the surface, reflection can be further reduced which increases the overall solar absorptance. New directions towards using machine learning to optimize the optical properties of thin film stacks [36] may enable the realization of selective solar absorbers with small footprint.

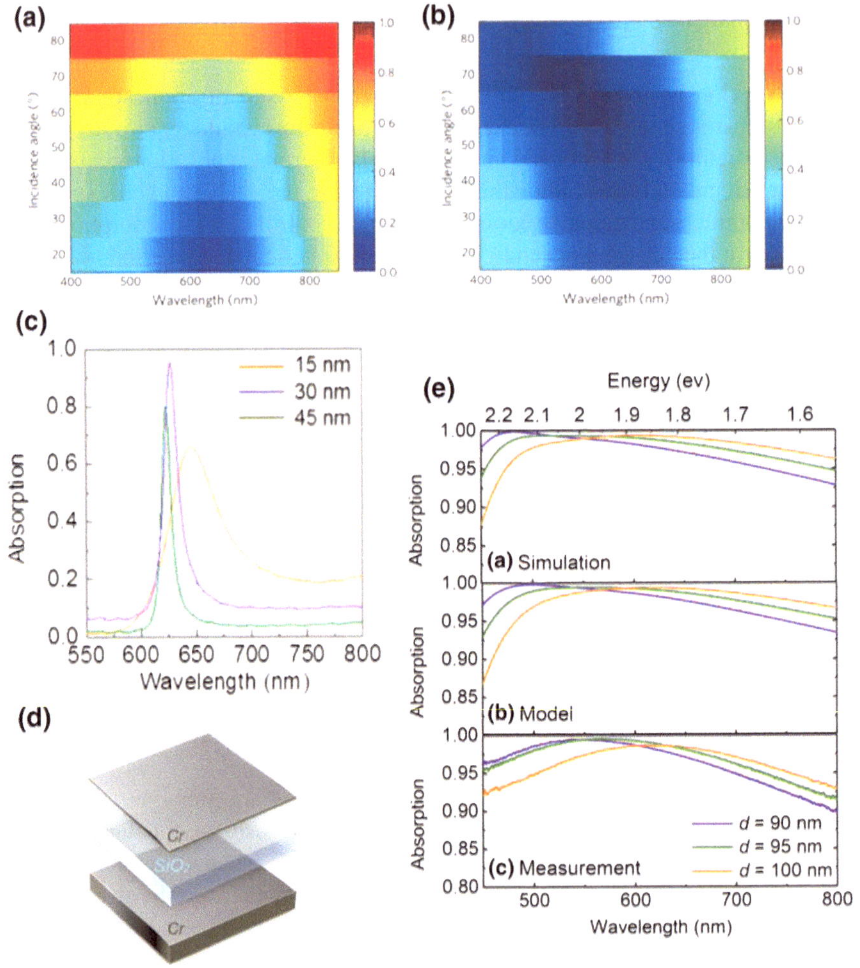

Fig. 1.5 Broadband light absorption for **a** s-polarized and **b** p-polarized light at different incidence angles. Reproduced with permission from Ref. [7], Springer Nature. **c** Shows the ability to tune the absorption bandwidth of an MDM Ag–SiO2–Ag cavity by changing the top Ag metal thickness. **d** Shows a schematic of a broadband absorber consisting of Cr–SiO_2–Cr MDM cavity. The cavity's absorption covers the entire visible range as shown in (**e**). (**c**), (**d**), and (**e**) are Reproduced with permission from Ref. [29]

1.4.3 Designer Profile of Optical Losses

For different light absorbers, optical losses occur at different layers of the thin films. For example, in an MDM cavity, losses, naturally, occur in the top and bottom metal films. On the other hand, for ultrathin lossy dielectric-high reflectance substrate, losses occur mainly inside the lossy dielectric while limited losses occur inside the reflective substrate (see Fig. 1.6a). The strong absorption inside the semiconductor is advantageous particularly for photodetectors and for photovoltaic applications. In fact, it was shown that more than 98% of the incident light energy can be absorbed in a 12 nm thick Ge layer on an Ag substrate at the wavelength of 625 nm over a wide range of angles using which exceeds the Yablonovitch limit, i.e., the limit that describes the maximum achievable light absorption of semiconductors in photovoltaic cells. This is because the Yablonovitch limit is not applicable if special nanophotonic trapping mechanisms are used [19]. Furthermore, by increasing the light absorption in an ultrathin semiconducting layer, this will naturally increase the efficiency of photodetectors. As we mentioned earlier, by using lossless dielectric on a high reflectance substrate and adding a lossy, ultrathin film as a superstrate, it is possible to locate the absorption within the ultrathin film which can have important applications in maximizing the emission intensity from ultrathin emitters, e.g., direct bandgap two-dimensional materials, and for optoelectronic, graphene-based applications.

Finally, in the case of light absorption based on lossless dielectric on a lossy substrate, the losses are exclusively located inside the absorptive substrate (see Fig. 1.6b). For metallic substrate, which has high thermal conductivity, the absorber more suitable for thermo-photovoltaics where the absorbed heat from solar spectrum is transferred to a selective emitter, as well as stealth technology, electromagnetic shielding since high thermal conductivity ensures better heat exchange and management. In addition, the ease of realizing strong light absorption in the IR with dielectric thickness $\ll\lambda$ makes it more suitable for selective thermal emission and radiative cooling.

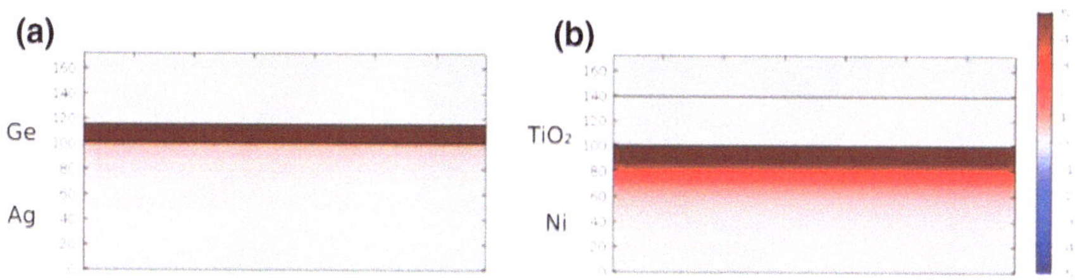

Fig. 1.6 Power dissipation density calculations for TE-polarized 650 nm wave normally incident on **a** a 20 nm Ge film on a 100 nm Ag substrate and **b** a 40 nm TiO_2 film on a 100 nm Ni substrate. Reproduced with permission from Ref. [3], John Wiley and Sons

1.4.4 Designer Perfect Light Absorption Angle

Controlling the angle where perfect light absorption occurs is another degree of freedom that can be of interest to certain applications where the absorber is used as a thermal emitter and the radiation angular range needs to be controlled. For example, for daytime and night-time radiative cooling via radiating within the atmospheric window, optimal performance is realized from emitters that emit normal to the surface and towards the sky since at larger angles, the atmosphere is thicker and is no longer transparent. According to Kirchhoff's law, $\alpha_\lambda(\theta) = \varepsilon_\lambda(\theta)$ where $\alpha_\lambda(\theta)$ and $\varepsilon_\lambda(\theta)$ are the absorptivity and emissivity at wavelength λ and direction θ, respectively. The thermal radiation intensity emitted from a body $\mathrm{I}(\lambda, \mathrm{T}, \theta) = \varepsilon_\lambda(\theta) E_b(\lambda, T)$, where and $E_b(\lambda, T)$ is the intensity of blackbody radiation at wavelength λ and temperature T determined by Planck's law. The absorptance thus determines the blackbody spectral radiance and characterizes the way a given body will emit thermal radiation [37].

The ability to control the angle where perfect light absorption takes place was studied in [3]. Figure 1.7a shows the absorptance of a Ni substrate and a dielectric superstrate with different values for n_d and t at normal incidence. A low n superstrate, for example, $n_d = 1.5$ cannot achieve strong light absorption regardless of its thickness which means that there is no strong light absorption whether the phase condition is satisfied or not since the amplitude condition cannot be satisfied. On the other hand, for a dielectric with moderately high n_d, for example, $n_d = 2.5$, the absorptance is higher and reaches strong light absorption at higher angles of incidence of TE polarized light as shown in Fig. 1.7b. This is because at higher angles the Fresnel reflection coefficient of dielectrics increases for TE polarization and, beyond 60°, for TM polarization. Alternatively, if the substrate has low reflectance, strong light absorption can be realized at low incidence angles even if the superstrate refractive index is low. Figure 1.7c shows the absorptance of normally incident light

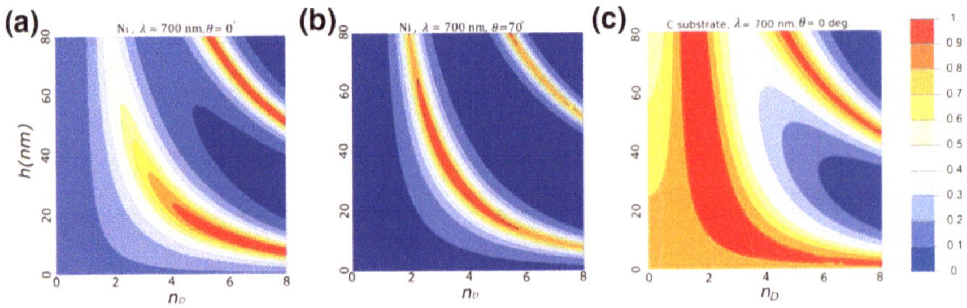

Fig. 1.7 **a** For a Ni substrate, the absorptance at normal incidence for a given lossless dielectric is shown as a function of its thickness h and its refractive index $\mathrm{n_d}$. For Ni, strong light absorption is not possible for $\mathrm{n_d} < 4$. However, at higher incidence angles strong to perfect light absorption is possible even for $\mathrm{n_d} \sim 2$ as shown in (**b**) due to the high reflection coefficient of dielectrics at higher angles. If perfect light absorption is desired at low incidence angles and low refractive index dielectric, one can use a substrate with low reflectance, e.g., carbon as shown in (**c**). Reproduced with permission from Ref. [3], John Wiley and Sons

($\lambda = 700$ nm) from air on an amorphous carbon substrate while varying n_d and t of the dielectric coating. Unlike the Ni substrate shown in Fig. 1.7a, strong light absorption is possible for a dielectric with low refractive index, and consequently reflectivity, with respect to light incident from air.

1.5 Iridescence Properties of Thin-Film Interference-Based Light Absorbers

Light absorption based on interference, naturally, will depend on the incidence angle, i.e., the absorption is iridescent. While some spectacular coloration in nature is due to iridescence and iridescent coloration is important for anti-counterfeiting applications, angular dependence may be problematic in many applications where a wide-angle range absorption is necessary, e.g., structural coloring for display technology, and electromagnetic shielding. The iridescence problem is not evident when it comes to broadband absorbers as they appear to be largely insensitive to the incident angle due to their broadband nature which results in a slight change in the absorption intensity at a given wavelength as a function of angle [33]. For narrowband absorbers, however, the angle dependence can be detrimental for practical applications, especially for structural coloring as their color would change significantly [29].

Omnidirectional absorption in MDM cavities was demonstrated by excitation of gap surface plasmon polariton (SPP) mode; a radiative surface plasmon wave that is excited at a metal-dielectric-metal interface when the dielectric layer is deeply subwavelength [38]. The demonstrated gap SPP absorption intensity, however, is reduced drastically for larger angles, which would result in effective color angle dependence. Furthermore, the excitation of Ferrel-Berreman (FB) mode enables iridescent free light absorption in a single thin film. This mode is a radiative mode, i.e., it couples directly from free space without any need of a momentum coupler and is excited near the frequency where the material's intrinsic permittivity goes to zero. The absorptive properties of this mode were shown in [39] and have been used to realize polarization switches [20] and to significantly enhance the nonlinear response of conductive metal oxide thin films [40]. However, this mode is only excited by TM mode and results in strong light absorption only at higher angles. Furthermore, metals exhibit this mode in the UV range [41], while conductive metal oxides exhibit the mode at the NIR-Mid IR range which excludes FB mode from many applications.

The iridescence of colors produced by thin films is abundant in nature and MDM cavities are not an exception as pointed out in [29]. A solution to it in the visible and NIR ranges was demonstrated in [27]. By adding a high index dielectric, the iridescence is reduced drastically. A high-index dielectric would make the refracted angle, inside the cavity, close in value even for different incidence angles. For the design of the NIR narrow band, iridescent free, perfect absorber, the authors used an Ag (13.5 nm)–Ge (130 nm)–Ag (70 nm) cavity. Note that the transmission is suppressed in the NIR even for a relatively thin Ag bottom layer, since Ag behaves as a nearly perfect electric conductor in the NIR. In order to realize perfect light absorption in the NIR, the Ag top layer should be ~10 nm. Having a very thin Ag film (<20 nm) can be problematic as thin Ag films tend to be rough and form islands. An irregular Ag film may cause unwanted scattering which can prevent the desired critical light coupling. This is another advantage for using Ge as a dielectric for the NIR absorber. Very thin Ag films deposited on top of Ge are homogeneous and avoid clustering which further increases light absorption due to reduced light scattering from surface roughness [42]. For p-polarized light, the authors realized perfect light absorption ($\approx$99.8%) at 1576 close to the telecommunication wavelength, with a Q-factor of 8 for 65° incidence angle.

Figure 1.8a shows the reflectivity spectrum for the Ag–Ge–Ag cavity from 15° to 75°. The absorption minimum remains almost unmodified as a function of the incident angle, while the absorption intensity slightly decreases at smaller angles. For comparison, an MDM cavity with low n_d dielectric MMA (8.5MMAEL 11, MICROCHEM). The MMA layer was spin-coated at 4000 rpm. Figure 1.8b shows

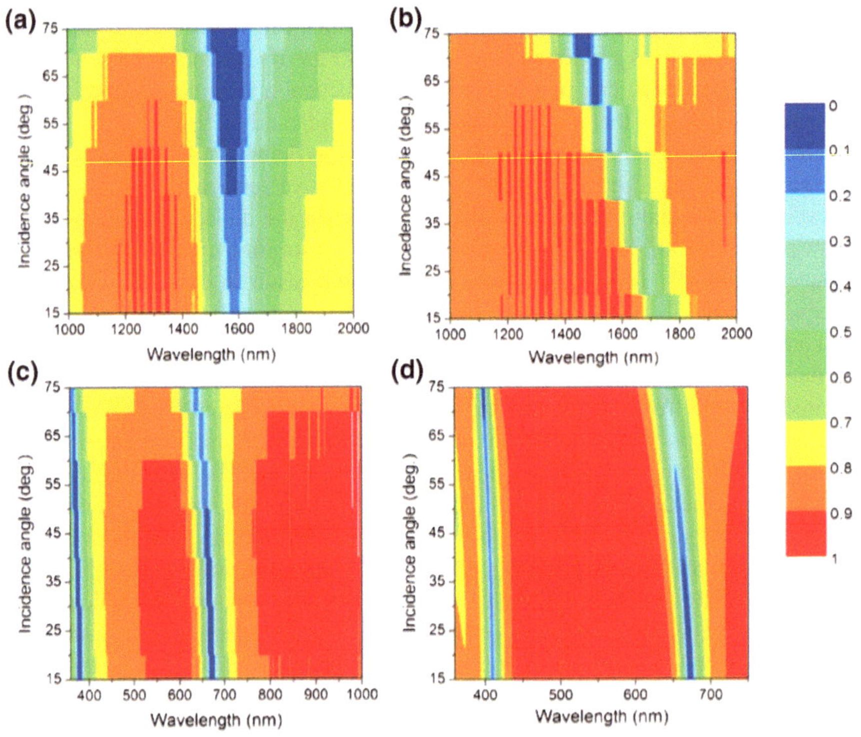

Fig. 1.8 **a** Angular reflectance of an Ag–Ge–Ag MDM cavity shows NIR iridescent free light absorption. Using a low refractive index dielectric, however, results in strong iridescence as shown in (**b**) for an Ag–MMA–Ag MDM cavity. **c** Measured, and **d** calculated angular reflectance for an Ag–TiO_2–Ag MDM cavity shows iridescent free light absorption in the visible for two cavity modes. Adapted with permission from Ref. [27], Optical Society of America

the reflectivity spectrum for an Ag (22 nm)–MMA (560 nm)–Ag (70 nm) cavity. The first-order mode (m = 1) of such a cavity coincides spectrally with the Ag–Ge–Ag cavity mode at 45° and suffers from considerable spectral shift in absorption for small angular changes. The reflection minimum for the MMA cavity shifts by ≈280 nm for the given angle range compared to only ≈48 nm shift for the Ge cavity. Using Ge as a narrowband absorber in the visible is not possible because it is highly absorbing due to direct electronic transitions at high photon energies [7]. On the other hand, TiO_2 has relatively high refractive index in the visible $n_d \approx 2.2$ and negligible losses. Figure 1.8c and d shows the measured and calculated reflectance spectrum for p-polarized light in the visible range from 360 to 750 nm and 15°–75° for Ag (30 nm)–TiO_2 (117 nm)–Ag (100 nm). The cavity shows two light absorption modes reaching ~94 and 96% absorption and Q-factors of ≈20 and ≈19, respectively, for a 15° incidence angle. Clearly, the modes are highly insensitive to the angle of incidence.

1.6 Thermally Induced Perfect Light Absorption in Low Reflectance Metals

Now that we discussed the angular dependence of light absorbers, we will take a detour back to discuss an old observation of a system that exhibit iridescent free, perfect light absorbers which was reexamined recently [43]. As we discussed previously, perfect light absorbers require a minimum of two separate material depositions, one for the metallic substrate and another for the dielectric film. On the other hand, it is known that heating reflective metals produces a wide range of colors due to oxidation of the metal surface. In fact, the most vibrant colors used in the pre-industrial era came from oxides, acetates and carbonates of metal ores and minerals. What is surprising, however, is that these colors were shown, in case of, Ni and Ti, are due to iridescent free, perfect light absorption that can be spectrally tuned even beyond the visible spectrum. Consequently, heating metallic thin films that create high index oxides can be used for other applications beyond creating vibrant colors. Moreover, the oxide layer forming on the metal has a thickness that is just few nanometers. As we showed earlier, a relatively high-index oxide on a moderately reflecting substrate can create strong to perfect light absorption which only require an oxide with thickness $t < \lambda/4n_d$.

The authors deposited 150 nm of Ni and 150 nm of Ti which creates an optically thick metallic base for the absorber, and is the only deposition required to build the perfect light absorber. After baking the samples at 400°C for 20–40 min in increments of 10 min for Ni and for 40–60 min for Ti, different colors started to appear. The reflectance spectra for heated Ni and Ti samples (incidence angle = 15°) are shown in Fig. 1.9. Clearly, unheated Ni and Ti films have weak absorptance across the visible spectrum (~50%) (Fig. 1.9a and b). The Ni film that was baked for 20 min absorbs 99.7% of light at 425 nm. As the baking time increased

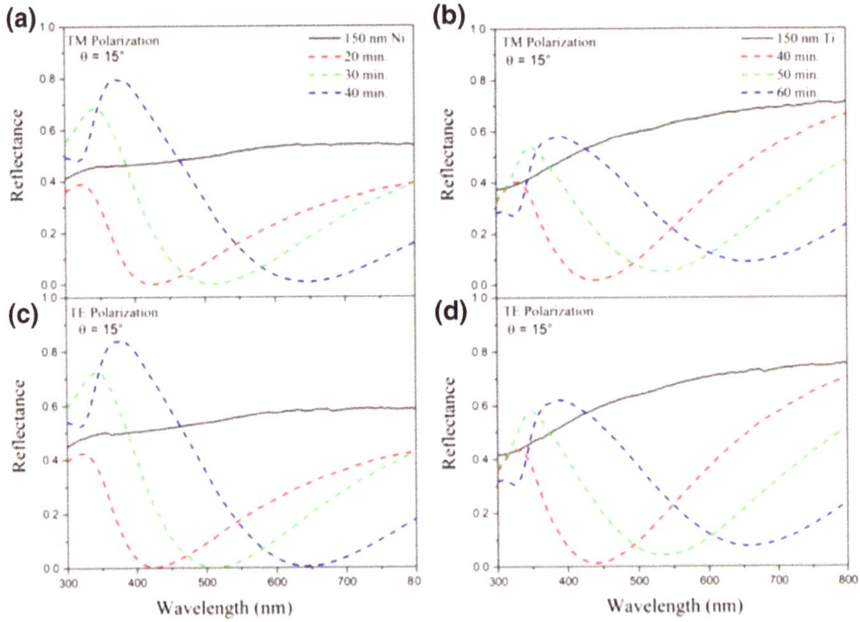

Fig. 1.9 Light absorption in heated metal films: The reflectance spectra for **a** TM, and **b** TE polarized light incident on an optically thick Ni film (black), the film heated for 20 min (red), 30 min (green) and 40 min (blue). Maximum absorbance of 99.93% is obtained. The reflectance spectra for **c** TM, and **d** TE polarized light incident on an optically thick Ti film (black), the film heated for 40 min (red), 50 min (green) and 60 min (blue). The absorptance is lower for heated Ti samples compared to Ni samples. For all measurements, the angle of incidence is 15°. Adapted with permission from Ref. [43], Optical Society of America

to 30 min, the resonance red shifts to 510 nm where 99.94% of light is absorbed. Baking for 40 min further redshifts the resonance to 645 nm and 99.5% of light is absorbed. Similarly, Fig. 1.9c and d show the reflectance spectra for an untreated Ti film as well as backed Ti films. Similar to Ni films, absorption resonance redshifts to longer wavelengths as the heating time is increased; however, perfect light absorption is not possible for the Ti–TiO_2 system at small angles. The oxide thickness was determined using spectroscopic ellipsometry measurements assuming a semi-infinite metallic substrate (either Ni or Ti) with a finite thickness of the associated metal oxide (NiO or TiO_2) on top using the transfer matrix method (TMM). The complex refractive index of NiO, $\tilde{n}_{NiO}$, and TiO_2, $\tilde{n}_{TiO_2}$ at 650 nm are $\tilde{n}_{NiO} = 3.27 - 0.09$ i, and $\tilde{n}_{TiO_2} = 2.34$. The NiO thickness for 20, 30 and 40 min baking duration are 12, 20, and 30 nm, respectively, which corresponds to an optical length at the reflection minimum $\left(\lambda_{min}^{NiO}\right)$ of $\lambda_{min}^{NiO}/11.5\tilde{n}_{NiO}$, $\lambda_{min}^{NiO}/8\tilde{n}_{NiO}$, $\lambda_{min}^{NiO}/6.6\tilde{n}_{NiO}$, respectively. The TiO_2 thickness for 40, 50 and 60 min baking duration are 18 nm, 36 nm, and 50 nm, respectively, which corresponds to an optical length at the reflection minimum $\left(\lambda_{min}^{TiO_2}\right)$ of $\lambda_{min}^{TiO_2}/10.5\tilde{n}_{TiO_2}$, $\lambda_{min}^{TiO_2}/5.8\tilde{n}_{TiO_2}$, $\lambda_{min}^{TiO_2}/5.5\tilde{n}_{TiO_2}$, respectively.

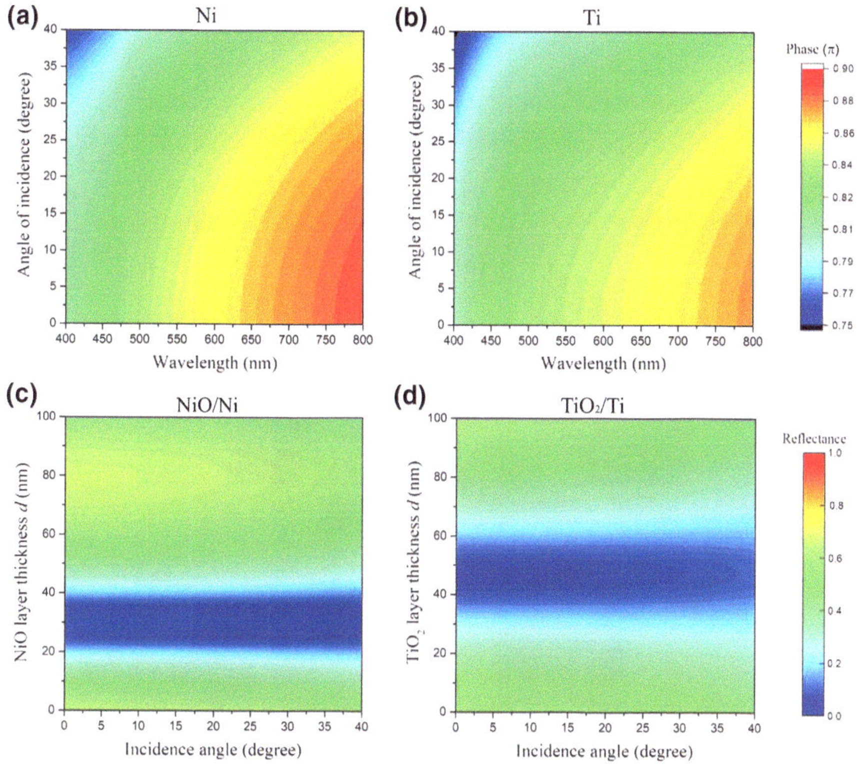

Fig. 1.10 Origin of ultrathin and iridescent free PLA. The calculated reflectance phase shift upon reflection of TM polarized light from **a** Ni, and **b** Ti, as a function of incidence angle and wavelength. The calculated angular reflectance as a function of incidence angle and **c** NiO thickness and **d** TiO_2 thickness reflects the iridescence-free absorption of the absorber. The incident wavelength is 650 nm. Adapted with permission from Ref. [43], Optical Society of America

The ultrathin oxides ($t < \lambda/4n_d$) are due to the nontrivial phase shift occurring at the oxide-metal interface. Figure 1.10a and b show the phase shift, in units of π, upon reflection from Ni and Ti at different angles of incidence (0°–40°) in the visible range. At longer wavelengths, the reflection phase approaches π as the metal reflectivity increases for lower frequencies, i.e., the metal approaches the perfect electric conductor condition. Furthermore, the high refractive index of the oxides means that very thin oxides can realize perfect light absorption. Figure 1.10c and d show the calculated reflectance of NiO/Ni and TiO_2/Ti for different NiO and TiO_2 thicknesses, respectively, as a function of incident angle at $\lambda = 650$ nm. Strong absorptance is achieved over wide angular range for both oxides. NiO absorptance, however, is less angle-sensitive, for a given oxide thickness, as it has higher refractive index [27].

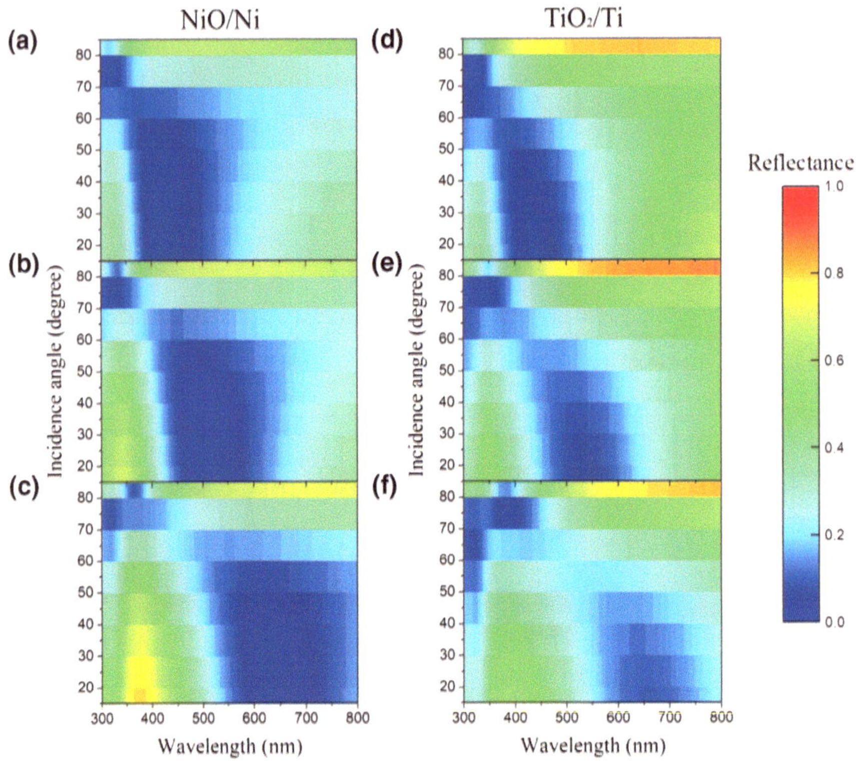

Fig. 1.11 Angular reflectance measurements: The angular reflectance from 15° to 85° for NiO–Ni absorber heated at 400 °C for **a** 20 min, **b** 30 min, **c** 40 min. The angular reflectance from 15° to 85° for TiO_2–Ti absorber heated at 400 °C for **d** 40 min, **e** 50 min, **f** 60 min. The angular insensitive of the absorptance is clear for both absorbers. The NiO–Ni has broader absorptance. Adapted with permission from Ref. [43], Optical Society of America

Figure 1.11a–c show the experimental angular reflectance of NiO/Ni, for NiO thicknesses of 12, 20, and 30 nm, respectively. Figure 1.11d–f show the experimental angular reflectance for TiO_2/Ti, for TiO_2 thicknesses of 18, 36, and 50 nm, respectively. The data presented are for TM polarized light, however, TE polarized light shows similar behavior. We note that the reflectance minimum for both samples is angle insensitive which explains the color persistence of oxidized metals at different angles. The broadness of the absorption modes for the NiO/Ni absorber is due to the existence of losses inside the weak sub-bandgap absorption in amorphous NiO layer.

1.7 Applications

1.7.1 Structural Colors Using Thin-Film Light Absorbers

Structural coloring; production of color by microscopically structured surfaces fine enough to interfere with visible light, is ubiquitous in nature and usually relies on light interference. Thin-film light absorbers have potential for structural coloring as they are a large-area, lithographically free method to create colors.

Reference [29] showed structural coloring using Ag–SiO_2–Ag transmission filters by combining deposition of thin films and FIB approach. Figure 1.12a shows a defined concentric circle pattern created with rainbow colors, in which the thickness of the dielectric layer is altered in a stepwise manner. Furthermore, a color palette consisting of 20 μm $\times$ 20 μm squares (Fig. 1.12b) with different dielectric thickness was fabricated which shows the ability to create micro images using this technique. Depending on the beam size of FIB, the resolution of such micro image could reach over 500 thousand dots per inch, which is roughly 1000$\times$ denser than any average smart phone display.

Note here, however, that similar colors are difficult to realize in the reflection mode since light absorbers create colors by subtraction, i.e., they reflect the complimentary color. This means that narrowband light absorbers would offer a Cayan-Magenta-Yellow colors and not Red-Green-Blue colors. Furthermore, color mixing is not possible when it comes to solid state structural coloration, at least when we adopt large area fabrication without resorting to FIB or other nanofabrication techniques. This problem was evident in this work [5]. The colors shown in Fig. 1.12a and b are iridescent. However, as shown in Fig. 1.12c–e iridescent free colors using an MDM cavity are possible by using a high index dielectric, here TiO_2 [27].

Broadband absorbers offer alternative colors, however, with limited variety as shown in Fig. 1.12f for lossy dielectric-high reflectance metal light absorber. Since absorption is localized mainly in the semiconductor, these absorbers can be used for decorative photovoltaic panels. Heating a low reflectance metal, as we discussed previously, result in PLA which translate into forming colors of some variety as shown in Fig. 1.12g. Using a lossless dielectric on a low reflectance substrate offers alternative, but limited, colors as shown in Fig. 1.12h. These colors are indeed iridescent free and, consequently, can be deposited on a rough surface. Figure 1.12i shows Ni film deposited on a roughened glass surface (frosted microscope glass slide). The ability to maintain the color integrity even on a roughened surface is due to the iridescent-free absorption.

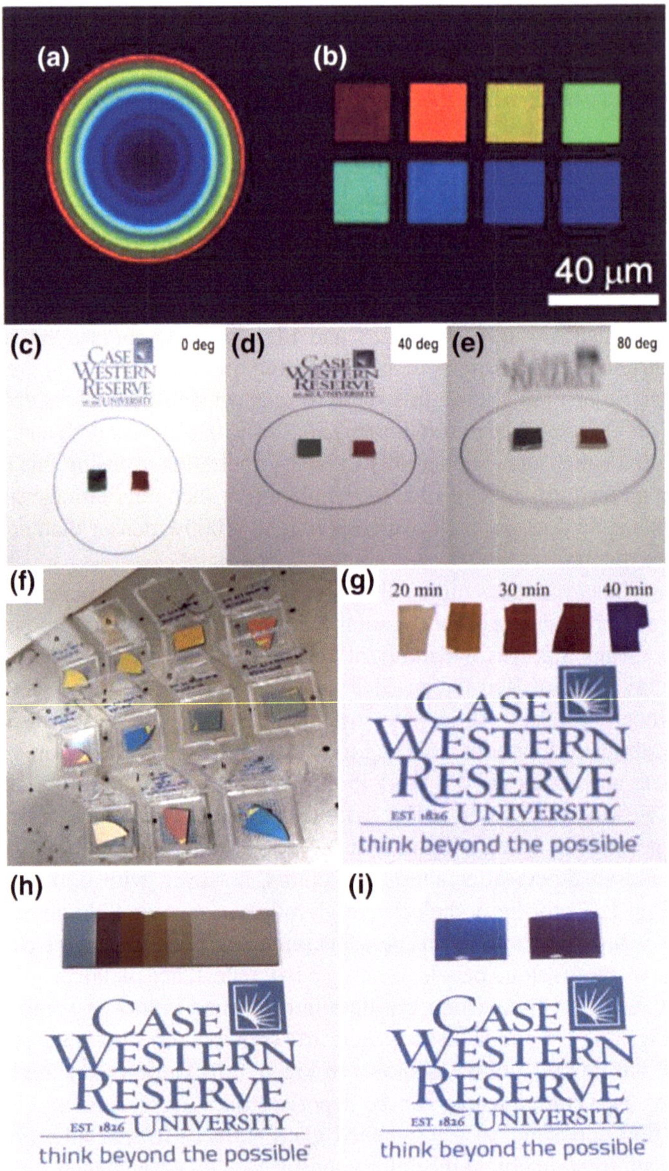
(a)
(b)
40 μm
(c)
0 deg
(d)
40 deg
(e)
80 deg
(f)
(g)
20 min
30 min
40 min
CASE WESTERN RESERVE
EST. 1826 UNIVERSITY
think beyond the possible
(h)
CASE WESTERN RESERVE
EST. 1826 UNIVERSITY
think beyond the possible
(i)
CASE WESTERN RESERVE
EST. 1826 UNIVERSITY
think beyond the possible

◂**Fig. 1.12** Photographs of **a** rainbow circles and **b** color pallets fabricated using focused ion beam (FIB) technique for patterning on the thin films with oxide thickness variation [29]. Reproduced with permission from Ref. [29], American Chemical Society. **c–e** Show photographs for two iridescent-free visible narrowband light absorbers are presented in for 0°, 40°, and 80°, respectively. Clearly, there is no color change for a wide-angle. Adapted with permission from Ref. [27], Optical Society of America. **f** Variety of colors formed by coating Au with nanometer films of Ge. a–h, 0, 3, 5, 7, 10, 15, 20 and 25 nm of Ge deposited on optically thick Au, which was deposited on polished silicon. i–k, 0, 10 and 20 nm of Ge deposited over 150 nm of Au, on unpolished silicon substrate. Reproduced with permission from Ref. [7], Springer Nature. **g** Colors produced on a 150 nm thick Ni film deposited on glass after heating it for 20–40 min. Reproduced with permission from Ref. [43]. **h** Photograph of a 100 nm thick Ni substrate with a TiO_2 film showing different colors for different TiO_2 thicknesses. **i** Depositing 100 nm thick Ni and TiO_2 film on frosted glass substrate still produce a single color due to the high refractive index of TiO_2. Reproduced with permission from Ref. [3], John Wiley and Sons

1.7.2 Hydrogen Gas Sensing Using Thin-Film Light Absorbers

At ambient pressure and temperature hydrogen is colorless, tasteless and highly flammable [44]. Hydrogen is flammable at concentrations ranging from 4 to 75% with low ignition energy [45]. Accordingly, hydrogen sensing is largely used in industries where it is a necessary component or a byproduct to monitor and control the hydrogen partial pressure for safety purposes. Hydrogen is also produced by certain bacteria; accordingly, hydrogen sensors are used in food industry and have possible medical applications [44]. While gas chromatography and mass spectrometry systems can be used in large scale industrial sensing, other applications require miniaturized sensors. Several nanophotonic and plasmonic systems were proposed as platform for optical hydrogen sensing [46, 47]. Photonic and plasmonic materials enable subwavelength control and manipulation of light and exhibit unique functionalities [48].

Metals that form metal hydrides, e.g. Palladium (Pd), upon exposure to hydrogen are excellent candidates for miniaturized hydrogen sensors due to their selective response to hydrogen. When Pd is exposed to hydrogen it forms palladium hydride (PdH_x) with certain hydrogen atomic ratio x. PdH_x exists in two solid state phases (α and β) depending on the temperature, pressure, and hydrogen partial pressure in its environment. At low hydrogen partial pressure, intercalated hydrogen atoms are randomly situated inside the Pd lattice and the hydrogen forms a solid solution (α phase) [49]. At higher hydrogen partial pressure, hydrogen-hydrogen interactions result in a more ordered structure forming a hydride β phase. In the β phase, the lattice parameter increases by ~3.5% as the lattice expands to accommodate the hydrogen atoms occupying new lattice sites [49]. For intermediate hydrogen partial pressures, both α and β phases coexist in a Pd film (so called α' phase). In α' and β phases, Pd experiences significant hysteresis between the formation and decomposition of the hydride phase. This hysteresis is because of the energy barrier between the hydride and metal phases due to lattice strains which does not exist in the α phase [45]. The change in the complex permittivity and lattice structure of

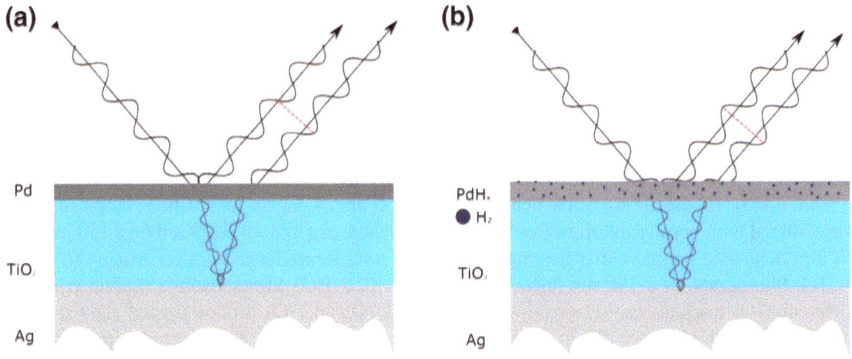

Fig. 1.13 Fabry-Perot MDM cavity hydrogen sensor: **a** A light absorber consisting of a metal (Pd)-dielectric (TiO_2)-metal (Ag) cavity. The light is absorbed due to destructive interference. **b** Exposing the absorber to hydrogen convert the Pd layer to a hydride PdH_x with a given hydrogen atomic ratio (x). The cavity can be designed such that formation of the hydride results in total destructive interference and perfect light absorption

Pd results in a measurable optical response from a photonic [50] or a plasmonic [47] system. For example, hydrogen sensing based on perfect light absorption was introduced in [48] where the authors used palladium nanowires as a grating on top of a multilayer system. In such system, reflection is reduced via impedance matching which results in perfect light absorption with absorptance $A \approx 99.5\%$. Changes in the optical properties in the palladium nanowire resulted in a shift in the absorption maximum $\Delta\lambda_{max} \approx 19$ nm, and a change in the measured relative intensity change of up to 500. Additionally, thin-film absorber hydrogen sensing was numerically investigated in [51]. While sensors based on metal hydrides are highly selective compared to other materials [52], these sensors are lithographically intense [47]. Realizing large area and lithography-free hydrogen sensor can overcome nanofabrication problems, e.g. high costs, low throughput and fabrication imperfection.

At resonance, the cavity absorption is maximized due to destructive interference [16]. The cavity resonance takes place when the phase difference between two successive reflected beams $\delta = \frac{4\pi}{\lambda_0} n' \, h \cos\theta' + \emptyset^t + \emptyset^b = 2m\pi$, where λ_0 and θ' are the incidence wavelength and angle respectively, n' and h are the refractive index and thickness of the dielectric forming the cavity respectively, $\emptyset^t$ and $\emptyset^b$ are the phase acquired upon reflecting from the top and bottom metal layers respectively and m is the resonance order and is an odd number. As shown schematically in Fig. 1.13a and b, transforming Pd to PdH_x changes $\emptyset^t$ and thus results in a shift in the resonant mode. Starting with a cavity that exhibits near-perfect light absorption, it is possible to choose an angle such that as Pd converts to PdH_x the absorption intensity increases by approaching an exact π phase difference between the partially reflected waves as shown in Fig. 1.13a and b.

The designed MDM sensor consists of 13 nm Pd–207 nm TiO_2–120 nm Ag which creates broadband light absorption. The reflectance for 0, 0.7, 2 and 4% H_2 exposure

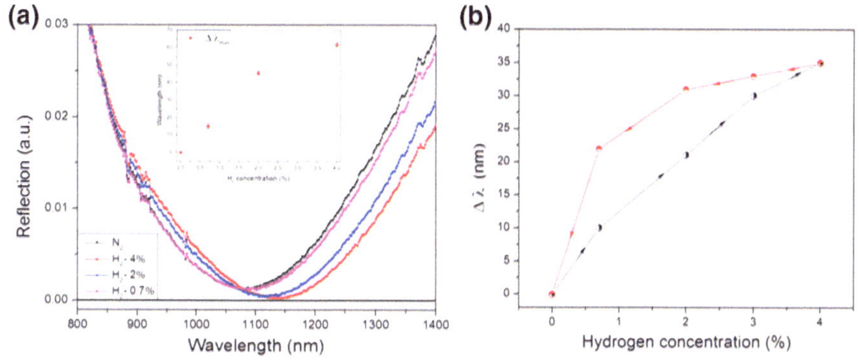

Fig. 1.14 **a** The reflectivity of the cavity absorber as a function of hydrogen concentration. The absorption mode red-shifts and decrease in absorption. The inset shows the λ_{min} as a function of concentration. **b** Hysteresis in MDM cavity absorber: Coherency strains affect the relaxation of the absorption cavity mode and result in hysteresis

are shown in Fig. 1.14a. The inset of Fig. 1.14 shows the measured shift in $\Delta\lambda_{min}$ for the different concentrations. The sensitivity $\Delta\lambda_{min}/\Delta C$ is $\sim 15 \pm 2$ nm per 1% increase in the hydrogen concentration. The total shift in λ_{min} at 4% H_2, $\Delta\lambda_{min} = 62$ nm compared to only 19 nm using palladium nanowire grating.

For hydrogen concentrations leading to the creation of α' or β phase, hysteresis in thin-film hydrogen sensors is always present. The hysteresis is due to the existence of a mechanical energy barrier due to the coherency strain produced by the lattice expansion in the hydride phase. The energy barrier prevents releasing hydrogen atoms situated inside the hydride lattice. Figure 1.14b shows the sensor response of the cycled hydrogen response for hydrogen concentrations (0, 0.7, 2, 3 and 4%). The hysteresis problem can be overcome, or at least attenuated, by introducing a metal hydride alloys [53].

References

1. Macleod HA (1986) Thin film optical filters, 4th edn. London Adam Hilger, London, UK
2. Khurgin JB (2015) How to deal with the loss in plasmonics and metamaterials. Nat Nanotech 10:2
3. ElKabbash M, Iram S, Letsou T, Hinczewski M, Strangi G (2018) Designer perfect light absorption using ultrathin lossless dielectrics on absorptive substrates. Adv Opt Mater 6:1800672
4. Watts CM, Liu X, Padilla WJ (2012) Metamaterial electromagnetic wave absorbers. Adv Mater 24:OP98
5. Cheng F, Gao J, Luk TS, Yang X (2015) Structural color printing based on plasmonic metasurfaces of perfect light absorption. Sci Rep 5:11045
6. Galinski H, Favraud G, Dong H, Gongora JST, Favaro G, Döbeli M, Spolenak R, Fratalocchi A, Capasso F (2017) Scalable, ultra-resistant structural colors based on network metamaterials. Light Sci Appl 6:e16233

7. Kats MA, Blanchard R, Genevet P, Capasso F (2012) Nanometre optical coatings based on strong interference effects in highly absorbing media. Nat Mater 12:20
8. Lee T, Jang J, Jeong H, Rho J (2018) Plasmonic- and dielectric-based structural coloring: from fundamentals to practical applications. Nano Converg 5:1
9. Wu Y-KR, Hollowell AE, Zhang C, Guo LJ (2013) Angle-insensitive structural colours based on metallic nanocavities and coloured pixels beyond the diffraction limit. Sci Rep 3:1194
10. Ju G et al (2015) High density heat-assisted magnetic recording media and advanced characterization—progress and challenges. IEEE Trans Magn 51:1
11. Huang X, El-Sayed MA (2011) Plasmonic photo-thermal therapy (PPTT). Alexandria J Med 47:1
12. Tian C, Qian W, Shao X, Xie Z, Cheng X, Liu S, Cheng Q, Liu B, Wang X (2016) Photoacoustic imaging: plasmonic nanoparticles with quantitatively controlled bioconjugation for photoacoustic imaging of live cancer cells. Adv Sci 3(12)
13. Bierman DM, Lenert A, Chan WR, Bhatia B, Celanović I, Soljačić M, Wang EN (2016) Enhanced photovoltaic energy conversion using thermally based spectral shaping. Nat Energy 1:16068
14. Kennedy CE (2002) Review of mid- to high-temperature solar selective absorber materials. NREL/TP-520-31267
15. Kraemer D, Jie Q, McEnaney K, Cao F, Liu W, Weinstein LA, Loomis J, Ren Z, Chen G (2016) Concentrating solar thermoelectric generators with a peak efficiency of 7.4%. Nat Energy 1:16153
16. Ni G, Li G, Boriskina Svetlana V, Li H, Yang W, Zhang T, Chen G (2016) Steam generation under one sun enabled by a floating structure with thermal concentration. Nat Energy 1:16126
17. Tao P, Ni G, Song C, Shang W, Wu J, Zhu J, Chen G, Deng T (2018) Solar-driven interfacial evaporation. Nat Energy 3:1031
18. Raman AP, Anoma MA, Zhu L, Rephaeli E Fan S (2014) Passive radiative cooling below ambient air temperature under direct sunlight. Nature 515:540
19. Kats MA, Capasso F (2016) Optical absorbers based on strong interference in ultra-thin films. Laser Photonics Rev 10:735
20. Yang Y, Kelley K, Sachet E, Campione S, Luk TS, Maria J-P, Sinclair MB, Brener I (2017) Femtosecond optical polarization switching using a cadmium oxide-based perfect absorber. Nat Photon 11:390
21. Chilwell J, Hodgkinson I (1984) Thin-films field-transfer matrix theory of planar multilayer waveguides and reflection from prism-loaded waveguides. J Opt Soc Am A 1:742
22. El Kabbash M, Rahimi Rashed A, Sreekanth KV, De Luca A, Infusino M, Strangi G (2016) Plasmon-exciton resonant energy transfer: across scales hybrid systems. J Nanomater 2016:4819040
23. ElKabbash M, Sousa-Castillo A, Nguyen Q, Mariño-Fernández R, Hoffman N, Correa-Duarte MA, Strangi G (2017) Tunable black gold: controlling the near-field coupling of immobilized Au nanoparticles embedded in mesoporous silica capsules. Adv Opt Mater 5:1700617
24. Cao L, White JS, Park J-S, Schuller JA, Clemens BM, Brongersma ML (2009) Engineering light absorption in semiconductor nanowire devices. Nat Mater 8:643
25. Kim SJ, Fan P, Kang J-H, Brongersma ML (2015) Creating semiconductor metafilms with designer absorption spectra. Nat Commun 6:7591
26. Landy NI, Sajuyigbe S, Mock JJ, Smith DR, Padilla WJ (2008) Perfect metamaterial absorber. Phys Rev Lett 100:207402
27. ElKabbash M, Ilker E, Letsou T, Hoffman N, Yaney A, Hinczewski M, Strangi G (2017) Iridescence-free and narrowband perfect light absorption in critically coupled metal high-index dielectric cavities. Opt Lett 42:3598
28. Tischler JR, Bradley MS, Bulović V (2006) Critically coupled resonators in vertical geometry using a planar mirror and a 5 nm thick absorbing film. Opt Lett 31:2045
29. Li Z, Butun S, Aydin K (2015) Large-area, lithography-free super absorbers and color filters at visible frequencies using ultrathin metallic films. ACS Photonics 2:183

30. Song H et al (2014) Nanocavity enhancement for ultra-thin film optical absorber. Adv Mater 26:2737
31. Jeong HY et al (2016) Optical gain in MoS2 via coupling with nanostructured substrate: Fabry-Perot interference and plasmonic excitation. ACS Nano 10:8192
32. Kats MA et al (2012) Ultra-thin perfect absorber employing a tunable phase change material. Appl Phys Lett 101:221101
33. Li Z, Palacios E, Butun S, Kocer H, Aydin K (2015) Omnidirectional, broadband light absorption using large-area, ultrathin lossy metallic film coatings. Sci Rep 5:15137
34. Bermel P et al (2012) Selective solar absorber. Annu Rev Heat Transfer 15:231
35. Li X-F, Chen Y-R, Miao J, Zhou P, Zheng Y-X, Chen L-Y, Lee Y-P (2007) High solar absorption of a multilayered thin film structure. Opt Exp 15:1907
36. Liu D, Tan Y, Khoram E, Yu Z (2018) Training deep neural networks for the inverse design of nanophotonic structures. ACS Photonics 5:1365
37. Campione S, Marquier F, Hugonin J-P, Ellis AR, Klem JF, Sinclair MB, Luk TS (2016) Directional and monochromatic thermal emitter from epsilon-near-zero conditions in semiconductor hyperbolic metamaterials. Sci Rep 6:34746
38. Shin H, Yanik MF, Fan S, Zia R, Brongersma ML (2004) Omnidirectional resonance in a metal–dielectric–metal geometry. Appl Phy Lett 84:4421
39. Luk TS et al (2014) Directional perfect absorption using deep subwavelength low-permittivity films. Phys Rev B 90:085411
40. Alam MZ, De Leon I, Boyd RW (2016) Large optical nonlinearity of indium tin oxide in its epsilon-near-zero region. Science 352:795
41. Newman WD, Cortes CL, Atkinson J, Pramanik S, DeCorby RG, Jacob Z (2015) Ferrell-Berreman modes in plasmonic epsilon-near-zero media. ACS Photonics 2:2
42. Logeeswaran VJ, Kobayashi NP, Islam MS, Wu W, Chaturvedi P, Fang NX, Wang SY, Williams RS (2009) Ultrasmooth silver thin films deposited with a germanium nucleation layer. Nano Lett 9:178
43. Letsou T, ElKabbash M, Iram S, Hinczewski M, Strangi G (2019) Heat-induced perfect light absorption in thin-film metasurfaces for structural coloring. Opt Mater Exp 9:1386
44. Wadell C, Syrenova S, Langhammer C (2014) Plasmonic hydrogen sensing with nanostructured metal hydrides. ACS Nano 8:11925
45. Fong NR, Berini P, Tait RN (2016) Hydrogen sensing with Pd-coated long-range surface plasmon membrane waveguides. Nanoscale 8:4284
46. Bévenot X, Trouillet A, Veillas C, Gagnaire H, Clément M (2002) Surface plasmon resonance hydrogen sensor using an optical fibre. Meas Sci Technol 13:118
47. Liu N, Tang ML, Hentschel M, Giessen H, Alivisatos AP (2011) Nanoantenna-enhanced gas sensing in a single tailored nanofocus. Nat Mater 10:631
48. Tittl A, Mai P, Taubert R, Dregely D, Liu N, Giessen H (2011) Palladium-based plasmonic perfect absorber in the visible wavelength range and its application to hydrogen sensing. Nano Lett 11:4366
49. Syrenova S et al (2015) Hydride formation thermodynamics and hysteresis in individual Pd nanocrystals with different size and shape. Nat Mater 14:1236
50. Zhao Y, Wu Q, Zhang Y (2017) High-sensitive hydrogen sensor based on photonic crystal fiber model interferometer. IEEE Trans Instrum Meas 66:2198
51. Serhatlioglu M, Ayas S, Biyikli N, Dana A, Solmaz ME (2016) Perfectly absorbing ultra thin interference coatings for hydrogen sensing. Opt Lett 41:1724
52. Ngene P, Radeva T, Slaman M, Westerwaal RJ, Schreuders H, Dam B (2014) Seeing hydrogen in colors: low-cost and highly sensitive eye readable hydrogen detectors. Adv Funct Mater 24:2374
53. Wadell C, Nugroho FAA, Lidström E, Iandolo B, Wagner JB, Langhammer C (2015) Hysteresis-free nanoplasmonic Pd–Au alloy hydrogen sensors. Nano Lett 15:3563

Chapter 2
Realization of Point-of-Darkness and Extreme Phase Singularity in Nanophotonic Cavities

The Aharonov–Bohm effect [1] and the Berry phase [2] are the two important physical phenomena that depends on non-trivial behavior of phase. In particular, singular photonics relies on the abrupt phase changes. The maximum phase change occurs at the point-of-darkness where the incident light is completely absorbed. This phenomenon has been widely investigated using different material systems through the concept of topological darkness [3]. However, to realize topological darkness, sub-wavelength nanostructures are required which demand intense nanofabrication steps. In this chapter, we demonstrate a new approach to realize point-of-darkness and extreme phase singularity using lithography-free metal-dielectric multilayer thin film stacks. We further demonstrate the potential applications of this concept in ultra-sensitive label-free biosensing.

2.1 Topological Darkness

It is well-known that the perfect absorption of light in optical structures is only possible when the combined reflection and transmission are completely suppressed. As discussed in Chap. 1, perfect absorption of light is an emerging research area that has been extensively studied in different spectral bands using various optical systems [4–7]. In optical systems, such as, for example, light reflection from a single interface [8], prism-coupled surface plasmon resonance [9], coherent absorption [10], and parity-time metamaterial [11], the perfect absorption (or zero reflection) of light can be achieved at certain values of the incidence angle for specific frequencies and polarization states. However, in fact, the complete suppression of reflection is not possible in such systems due to sample fabrication errors such as disorder, inhomogeneity, etc.

In order to overcome these technological hurdles and push the limits of complete suppression of reflection, topological darkness concept has been proposed by Kravets et al. [3]. An optical system that follows the Jordan Curve Theorem [12] can exactly provide complete suppression of reflection at certain incidence angles and

Sreekanth K. V. et al., *New Directions in Thin Film Nanophotonics*, Progress in Optical Science and Photonics 6, https://doi.org/10.1007/978-981-13-8891-0_2

frequencies. In particular, topological darkness is a phenomenon described in the two-dimensional optical-constant (n, k) plane with a slight restriction on the effective dispersion curve [n_{eff} (λ), k_{eff} (λ)] such that the dispersion curve must start and end in two different areas separated by the zero-reflection curve. It shows that zero reflection point is topologically protected such that disorder of the samples cannot be affected. The abrupt phase change at the point-of-darkness has been demonstrated using various optical systems [3, 13–16]. A gold nanorod-based plasmonic metamaterial was first proposed to realize topological darkness in the visible wavelengths [3]. Later, Malassis et al. have been used self-assembled plasmonic metamaterials to realize topological darkness [13]. A 3D plasmonic crystal metamaterial consisting of metallized woodpile structure has been proposed for realizing topological darkness [14]. Recently, dispersion topological darkness at multiple wavelengths and polarization states using a three-layered structure with nanopatterned top layer was experimentally demonstrated [15]. However, to date, abrupt phase change at the point-of-darkness has only been demonstrated using sub-wavelength nanostructures so that the effective dispersion curve can be engineered. In addition, complex nanofabrication techniques are required to develop sub-wavelength nanostructures, especially at visible frequencies.

2.2 Lithography-Free Nanophotonic Cavities

In this section, we propose a new route to achieve abrupt phase change at the point-of-darkness using a planar four-layered metal-dielectric-dielectric-metal stack [17]. In the proposed asymmetric Fabry-Perot (F-P) cavity system, the point-of-darkness and singular phase are achieved due to the presence of a highly absorbing ultrathin layer of germanium (Ge) within the stack. As experimentally demonstrated by Kats et al. [18], an ultrathin Ge layer with a thickness much smaller than the illumination wavelength deposited on a finite optical conductivity thin metal film rapidly attenuates the incident light. In this pioneer work, authors have demonstrated that a strong interference effect occurs when a highly absorbing thin film of Ge on an optically thick Au film. In fact, the non-trivial phase shift occurs at the interface between Ge and Au leads to a strong resonant behavior.

The proposed four-layered asymmetric structure deposited on a silicon substrate is shown in the inset of Fig. 2.1c. It consists of a top and bottom Ag layer with thickness 20 and 80 nm, respectively. An optically thick (522 nm) lossless methyl methacrylate (MMA) layer as the cavity layer. A key component of the structure is the highly absorbing ultrathin layer of Ge with thickness 10 nm, which is deposited between MMA and bottom Ag layers. In the fabrication process, this films of Ag and Ge were deposited by thermal evaporation of Ag and Ge pellets at a deposition rate of 0.2 and 0.1 A/s, respectively. The optically thick layer of MMA was spin-coated on the samples. The spectroscopic ellipsometer was used to obtain the optical constants and thicknesses of all thin films. By performing field emission scanning

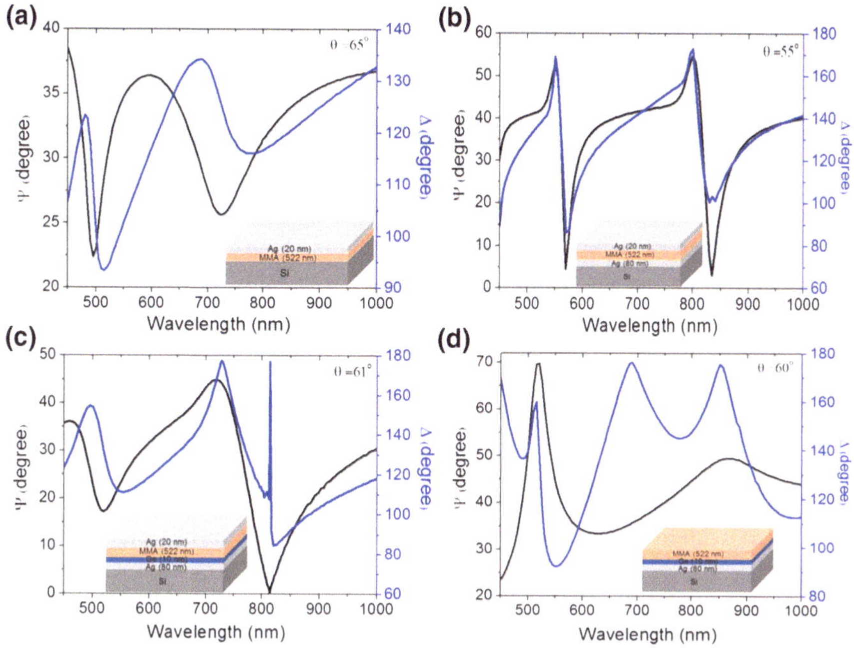

Fig. 2.1 Measured ellipsometry parameters (ψ and Δ) of a **a** Ag/MMA system, **b** Ag/MMA/Ag system, **c** Ag/MMA/Ge/Ag system, and **d** MMA/Ge/Ag system. All spectra were recorded for an incidence angle at which minimum/maximum ψ was obtained for the mode at longer wavelength. Reproduced with permission from Ref. [17]

electron microscopy (FE-SEM) characterization, we conformed that the thermally evaporated 10 nm thick Ge film has a uniform surface texture and morphology.

As it is known, ellipsometry measures the complex reflectance of a system, ρ which is parametrized by the amplitude component ψ and the phase difference Δ. Accordingly, ellipsometry parameters such as ψ and Δ have unique characteristics at the Brewster angle. In particular, ψ reaches a minimum/maximum value at the Brewster angle. In order to demonstrate point-of-darkness and singular phase at the Brewster angle, we recorded a pair of ellipsometry parameters (ψ, Δ) as a function of wavelength (400–1000 nm) and angle of incidence (20°–90°). In the ellipsometry measurements, ψ can vary from 0 to 90° and Δ ranges from 0 to 360° (or −180° to +180°). We fabricated different control samples and analyzed their data with that of the proposed sample. In Fig. 2.1, we show the measured ellipsometry parameters (ψ and Δ) of the fabricated four samples as a function of wavelength. The spectra were recorded for the incidence angle at which minimum/maximum ψ was obtained for the mode at longer wavelength.

The spectra of a two-layered (Ag/MMA) system is plotted in Fig. 2.1a, which supports two cavity modes one at 490 nm and another at 720 nm. By depositing an 80 nm thick bottom Ag film, these two modes are redshifted and the quality factor of the modes improved significantly (Fig. 2.1b). In particular, the ψ value of the second mode reached a minimum value of 3° at 835 nm wavelength and 55° angle of incidence. We also noticed that both modes showed perfect absorption for a wide range of angles of incidence. However, no singular phase behavior is observed for any combination of the thicknesses of MMA and bottom Ag layers for Ag/MMA/Ag system.

In Fig. 2.1c, we show the key results of our findings. In the case of the proposed four-layered system, both modes are blue shifted in comparison to Fig. 2.1b. It is important to note that the minimum ψ value corresponding to longer wavelength mode reached almost zero upon the deposition of a thin Ge layer between MMA and the bottom Ag film. More importantly, in contrast to Fig. 2.1b, an abrupt phase jump is realized at the reflection-less point where $\psi \approx 0$. It indicates that the singular phase at the point-of-darkness is obtained at a wavelength of 814 nm and an angle of incidence of 61°. We then removed the top Ag layer from the four-layered system and the spectra obtained at 60° appeared to be totally different, which is shown in Fig. 2.1d. In particular, the ψ spectrum of the MMA/Ge/Ag system shows a reverse behavior. It is due to the fact that the s-polarized light provides the minimum reflection at the resonance wavelength and incidence angle since $\tan \Psi = \left|\frac{r_p}{r_s}\right|$. However, p-polarized light provides the minimum reflection for other three samples (we have detailed this in the following sections). We further studied MMA/Ge/Ag system and confirmed that the generalization of Brewster angle is possible for this system, where the reflection of both p- and s-polarized light becomes zero at different angles of incidence and wavelengths. In short, the two modes observed in the case of all four samples are the cavity modes of the Ag/MMA system, however, their behavior can be altered by the deposition of bottom thin films of Ag and Ge with appropriate thicknesses as well as by removing the top Ag layer.

2.3 Singular Phase at the Point-of-Darkness

As shown in Fig. 2.1c, only the mode at longer wavelength demonstrates the point-of-darkness and singular phase, therefore we focus on this mode in the following discussions. It is clear that the physical mechanism of achieving a singular phase in the Ag/MMA/Ge/Ag system is uniquely due to the presence of an ultrathin highly absorbing Ge layer and thus it is a special case of an asymmetric F-P absorber cavity. In fact, since the thickness of Ge layer is much smaller than the operating wavelength, strong absorption resonance occurs. That is, the thickness of Ge, $d \approx \lambda/(20n)$ at $\lambda = 814$ nm and $n = 4.2$ (the measured refractive index of Ge at 814 nm). The measured ellipsometry spectra in wavelength and angular scan is shown in Fig. 2.2a, c, respectively. One can see that $\psi_{\min}$ and the phase singularity are obtained exactly at 814 nm and 61°. In contrast to wavelength scan, the lowest $\psi_{\min}$ is recorded in angular scan, therefore a higher phase change is achieved at the point-of-darkness. However, the measured total phase changes close to point-of-darkness (above and

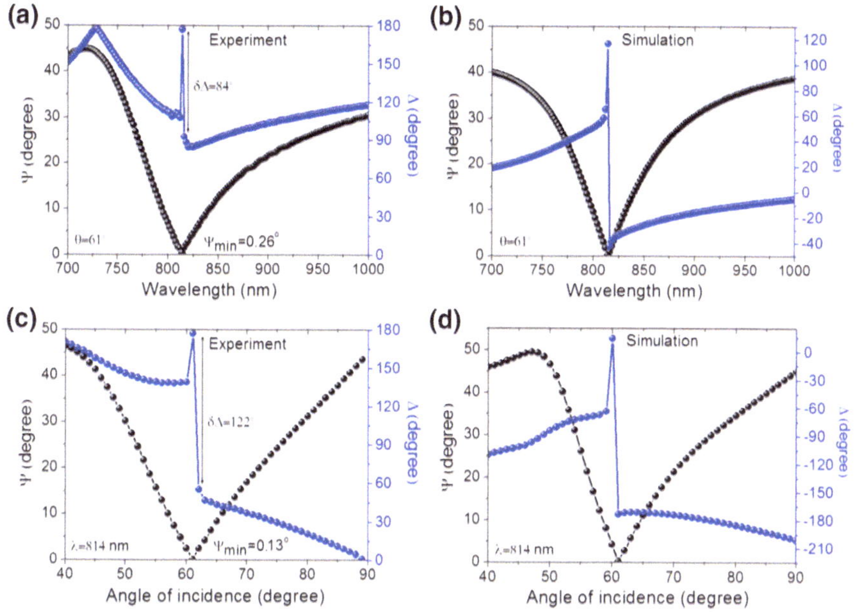

Fig. 2.2 Point-of-darkness and singular phase for the mode at the longer wavelength. In wavelength scan **a** measured and **b** calculated pair of ellipsometry parameters at an angle of incidence of 61°. In angular scan, **c** measured and **d** calculated pair of ellipsometry parameters at a wavelength of 814 nm. Reproduced with permission from Ref. [17]

below ψ_{min}) is almost the same in both scans. According to ellipsometry parameters, a general formula for calculating the figure of merit (FOM) of the system is given by $\delta\Delta/\delta\psi$, where $\delta\Delta$ and $\delta\psi$ is the change in the phase and amplitude, respectively [3]. This formula can be simplified as, FOM $= 1/\psi_{min}$, where ψ_{min} is in radians. The experimentally determined FOM of the proposed system is 445, where $\psi_{min} = 0.13°$ at 61°. However, the maximum FOM reported previously was 200, which was achieved using topological darkness-based singular phase concept [3].

We then solved Fresnel's equations for an *N*-layer model, using the transfer matrix method (TMM) to simulate the pair of ellipsometry parameters of the proposed system [17]. In this approach, we consider Fresnel's reflection coefficients ($r_{p,s}$) are a serial product of the interface matrix I_{jk} ($j = 0, 1, 2, 3 \ldots$ and $k = j + 1$) and the layer matrix L_j.

That is,

$$r_{p,s} = \frac{M_{12}}{M_{22}}, \quad \text{with } M = \begin{bmatrix} M_{11} & M_{12} \\ M_{21} & M_{22} \end{bmatrix} \tag{2.1}$$

$$I_{jk} = \begin{bmatrix} 1 & r_{jk} \\ r_{jk} & 1 \end{bmatrix} \text{ and } L_j = \begin{bmatrix} e^{ik_{zj}d_j} & 0 \\ 0 & e^{-ik_{zj}d_j} \end{bmatrix} \tag{2.2}$$

In Eq. (2.2), d_j is the thickness of the jth layer, and $k_{zj} = \sqrt{\varepsilon_j\left(\frac{2\pi}{\lambda}\right)^2 - k_x^2}$, with k_x is the parallel wave vector. The Fresnel's equations for p-and s-polarization are given by,

$$\text{For } p\text{-polarization,} \quad r_{jk} = \frac{\left(\frac{k_{zj}}{\varepsilon_j} - \frac{k_{zk}}{\varepsilon_k}\right)}{\left(\frac{k_{zj}}{\varepsilon_j} + \frac{k_{zk}}{\varepsilon_k}\right)} \tag{2.3}$$

$$\text{For } s\text{-polarization,} \quad r_{jk} = \frac{\left(\frac{k_{zj}}{\sqrt{\varepsilon_j\varepsilon_k}} - \frac{k_{zk}}{\sqrt{\varepsilon_j\varepsilon_k}}\right)}{\left(\frac{k_{zj}}{\sqrt{\varepsilon_j\varepsilon_k}} + \frac{k_{zk}}{\sqrt{\varepsilon_j\varepsilon_k}}\right)} \tag{2.4}$$

The ellipsometry parameter ψ is obtained from $\tan\Psi = \left|\frac{r_p}{r_s}\right|$, and the phase difference between the s- and p-polarization represents, $\Delta = \phi_s - \phi_p$, where the values of ϕ_s and ϕ_p are extracted from the Fresnel's reflection coefficients of the s- and p-polarization, $r_s = |r_s|\exp(i\phi_s)$ and $r_p = |r_p|\exp(i\phi_p)$, respectively. The simulated ψ and Δ spectrum of the proposed cavity in wavelength and angular scan is presented in Fig. 2.2b, d, respectively. As can be seen, the calculated results qualitatively match with experimental results. Note that the ellipsometrically acquired and fitted optical constants of Ag, Ge, MMA and Si were used in the simulations [17]. In order to obtain a good fit, the layer thicknesses used in Fig. 2.2b are 21, 538, 11 and 80 nm for the top Ag, MMA, Ge and bottom Ag, respectively, whereas the corresponding thicknesses used in Fig. 2.2d are 18, 532, 9 and 80 nm. We further verified the variance in phase change at the point-of-darkness by changing the thickness of Ge and Ag layers. Indeed, the abrupt phase change at the reflection-less point is extremely sensitive to the thickness of Ge layer.

In Brewster angle microscopy, it is a known fact that the p-polarized reflected intensity becomes zero when the light is incident on a semi-infinite transparent medium. However, this condition violates if the medium is an absorbing film such as metal or semiconductor with complex refractive indices. Even though, in principle, the Brewster angle microscopy condition does not exist in the case of the proposed system as it consists of absorbing media, we show that singular phase is obtained at the Brewster angle. To verify this, we recorded the reflectance spectrum for both p- and s-polarizations.

For p- and s-polarization, the measured and calculated ($R_{p,s} = |r_{p,s}|^2$) reflectance spectra of the four-layered system at 61° is shown in Fig. 2.3a, b, respectively. It is evident that p-polarized reflectance at 814 nm provides exactly zero reflectance, which confirms that the incidence angle (61°) at which zero reflectance obtained is the Brewster angle. Even though, almost zero reflection is obtained for Ag/MMA/Ag at p-polarization, the introduction of an ultrathin Ge layer into the three-layered system further decrease the p-polarized reflection and increase the s-polarized reflection, as a result almost zero ψ and abrupt phase change is obtained at the Brewster angle for Ag/MMA/Ge/Ag system [17]. To show the important role of Ge in the proposed system, we simulated intensity distribution along the mul-

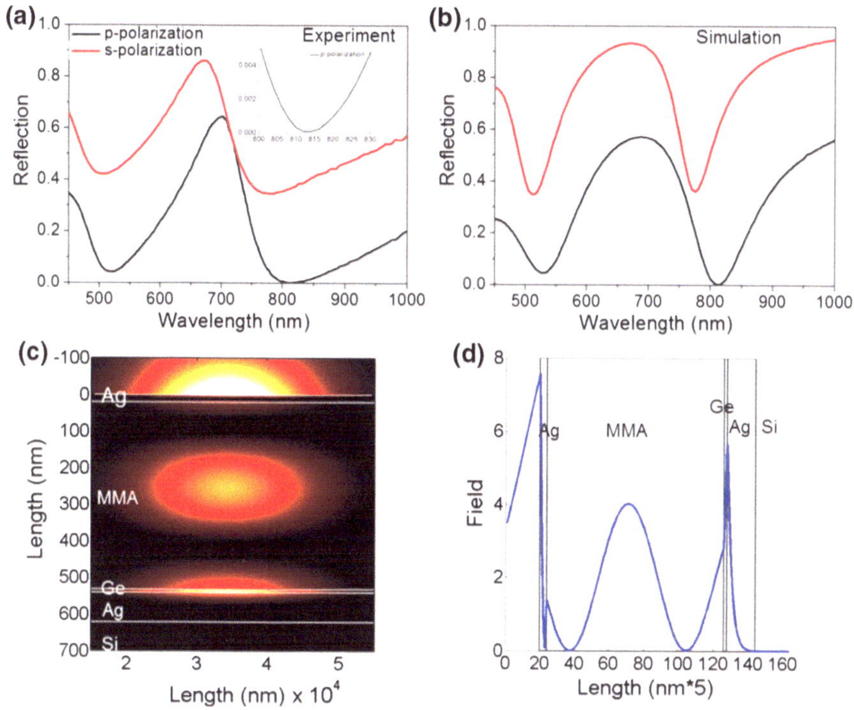

Fig. 2.3 Reflection spectrum of *p*- and *s*-polarization at 61°. **A** Measured and **b** calculated. The measured reflected intensity for *p*-polarization at 814 nm is 0.001%. **c** Simulated intensity distribution along the four-layered system at 814 nm and 61° and **d** its magnitude as a function of depth along the multilayer. Reproduced with permission from Ref. [17]

tilayer using TMM. Intensity distribution obtained at 814 nm and 61° is shown in Fig. 2.3c for 2D map and the corresponding line profile is shown in Fig. 2.3d. It is clear that the field intensity is tightly confined inside the ultrathin Ge layer at the point-of-darkness. This result further indicates as a direct evidence of the physical mechanism for achieving the point-of-darkness where the ultrathin layer of Ge heavily attenuates the incident light.

2.4 Phase-Sensitive Biosensing

Optical transduction methods have received considerable attention for the real-time monitoring of biomolecular binding events since they avoid the time-consuming labeling process [19]. Various label-free refractometric optical sensors based on different interrogation schemes such as wavelength, angle and intensity have been demonstrated. Among these, plasmonic nanostructures [20], metamaterials [21–23], photonic crystals [24] and whispering gallery mode-based techniques [25] showed better sensitivity for the detection of small analyte volumes. In recent years, phase-sensitive interrogation schemes [26, 27] have demonstrated higher sensitivity compared to sensors based on spectroscopic techniques. One of the most important applications of topological darkness demonstrated so far is the phase-sensitive optical biosensing because a singular behavior of the phase in Fourier space is possible at the point-of-darkness [3]. Notably, it is possible to monitor slight changes in the electromagnetic environment by measuring the phase changes at the topological darkness condition. However, it is not straightforward to use the topological darkness ($\psi = 0$) for sensing purpose due to the absence of light and abrupt phase jump. In this scenario, close to zero reflection condition ($\psi \approx 0$) is a requirement for biosensing, which can be achieved with ligand functionalization. Singular phase-based biosensing has been demonstrated using plasmonic gold nanoarray metamaterial, where an areal mass sensitivity at a level of $fgmm^{-1}$ with single molecule sensitivity has been realized [3]. Such an extreme sensitivity through phase interrogation scheme was achieved by diffractive coupling of the localized plasmon resonances of plasmonic metamaterial. In a work by Svedendahl et al. [16], the reflection-less condition attained only through localized surface plasmon resonances also demonstrated much higher phase sensitivity. However, all the optical systems that attained zero-reflection through topological darkness are based on sub-wavelength nanostructures, which demand complex nanopatterning techniques and restricts the scalability of the samples.

To overcome these issues, we propose the four-layered cavity as an alternative potential choice for the development of cost-effective phase-sensitive label-free refractometric biosensors. The operating principle of the sensor is based on the fact that a strong decaying field is possible along the superstrate since the top layer is a thin Ag film (see Fig. 2.3c). Initially, we develop a proof-of-concept biosensor platform and demonstrate the detection of streptavidin (molecular weight: 52.8 kDa) proteins by immobilizing the Ag surface with an ultrathin layer of small protein such as biotin (molecular weight: 244 Da).

In the immobilization process of biotin: we first prepared biotin disulfide using a modified synthetic protocol and a proper biotin-thiol was prepared for the immobilization of Ag surface [17]. We conducted X-ray photoelectron spectroscopy analysis to confirm the surface functionalization, where we traced the elemental sulfur and nitrogen residues post-modification. The fabricated cavities were immersed in different concentrations (0.1, 1 and 10 mM) of biotin-thiol prepared in phosphate-buffered-saline (PBS) buffer overnight. Later, the immobilized cavities were washed with PBS to remove the weakly attached and unbound biotin-thiol. Then, as such cavities were used in the sensing experiments. In order to inject streptavidin solutions on the sensor surface, we used a detachable polydimethylsiloxane (PDMS) channel and removed the channel while doing measurements because no singular phase is visible when channel is present.

In wavelength scan, the measured ψ and Δ spectrum of the sample by covering 1 mM biotin is shown as the black curve in Fig. 2.4a, b, respectively. As can be seen, the minimum ψ is obtained at slightly higher incidence angle of 65°, as a

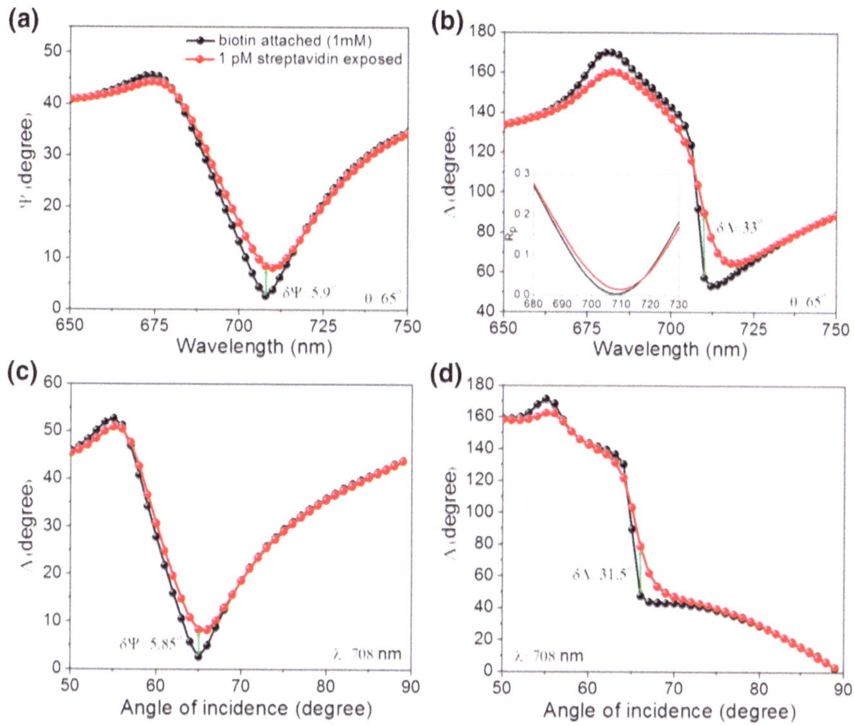

Fig. 2.4 In wavelength scan, measured ellipsometry parameters at 65°. **a** ψ and **b** Δ before (black curve) and after (red curve) exposure to 1 pM streptavidin. In angular scan, measured ellipsometry parameters at 708 nm **c** ψ and **d** Δ before (black curve) and after (red curve) exposure to 1 pM streptavidin. Reproduced with permission from Ref. [17]

result the wavelength corresponding to ψ_{min} is drastically blue shifted to 708 nm. The observed large blue shift is also due to the increase in the surface roughness of biotin immobilized sample. The measured ψ and Δ spectrum in angular scan by fixing the excitation wavelength at 708 nm is shown as the black curve in Fig. 2.4c, d, respectively. In order to demonstrate the detection of streptavidin, biotin covered samples were exposed to 1 pM streptavidin prepared in PBS for 2 h, followed by rinsing with PBS to remove the non-specifically bound streptavidin molecules. Note that this procedure was performed without removing the sample from the ellipsometer stage to keep the same incident beam position on the sample so that we could extract the exact shift.

After exposure to 1 pM streptavidin, the ellipsometry parameters (ψ and Δ) recorded in wavelength and angular scans are shown as the red curve in Fig. 2.4a–d, respectively. In the inset of Fig. 2.4b, the *p*-polarized reflectance spectrum recorded before and after exposure of streptavidin are also shown. In both wavelength and angular scans, a significant phase shift is obtained at the ψ_{min} value as compared to the corresponding ψ change. Notably, a maximum Δ change (δΔ) of 33° is recorded at 65° for 1 pM streptavidin concentration. We further studied the specific adsorption of streptavidin on the sensor surface by conducting an experiment in which the results of different concentrations of biotin attached samples were compared and confirmed that the obtained phase sensitivity is due to the refractive index changes caused by the specific binding of streptavidin.

By considering the illuminated beam diameter on the sample (~1 mm^2) and the dimension of the PDMS channel (7 × 7 × 2 mm^3), the streptavidin molecule (N_{max}) concentration at 1 pM concentration is estimated, which is around 11.8×10^5. Thus, the calculated device sensitivity in terms of phase shift ($\delta\Delta/N_{max}$) is $2.8 \times 10^{-5\circ}$, which is one order of magnitude higher than the obtained spectral sensitivity, $\delta\lambda/N_{max}$. The phase sensitivity of device can be further improved by narrowing the linewidth of the cavity mode, as a result high-quality factor mode can be realized.

2.5 Microfluidics Integrated Cavities

As mentioned above, we removed the PDMS channel while doing the sensing experiments because it is not possible to realize singular phase in reflection mode using a PDMS channel integrated cavity. It shows that singular phase highly depends on the topology of the cavity surface. In order to solve this critical issue particularly for the monitoring of real-time binding events, we designed and integrated a specially designed poly methyl methacrylate (PMMA)-based microfluidic channel with the cavity [28]. In Fig. 2.5a, b, we respectively show a schematic representation and photograph of the developed PMMA microfluidic channel integrated cavity. In contrast to the cavity design shown in Fig. 2.1c, the top metal layer is replaced with Au because the biomolecules can easily be immobilized with Au using a well-known thiol-based surface chemistry. Here, the thin film of Au was deposited by thermal evaporation of Au pellets at a deposition rate of 0.2 Å/s and a base pressure of $<5 \times 10^{-6}$ mbar.

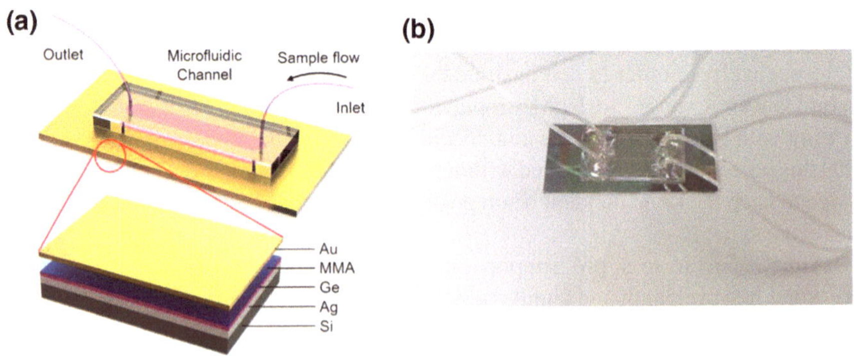

Fig. 2.5 **a** A schematic representation of microfluidic channel integrated cavity. The thickness of Au, MMA, Ge and Ag layer is 20, 520, 10 and 80 nm, respectively. **b** A photograph of the sensor device with microfluidic channel and sample tubing. Reproduced with permission from Ref. [28]

A key component of the sensor device is the PMMA-based microfluidic channel consisting of a PMMA plastic top (encompassing micromachined inlets and outlets) and a double-sided adhesive film with a thickness of 50 μm [28]. The adhesive film defines the outlines and thickness of the microfluidic channel. By laser micromachining, PMMA tops are fabricated with an inlet and outlet with a diameter 0.61 mm and a separation of 14 mm. Optically clear double-sided adhesive tape is laser micromachined to encompass the 16 mm × 3 mm microfluidic channel with 50 μm height. The micromachined tape is then attached to the PMMA component to include the inlet and outlet between the outline of the channels. To form microfluidic channel integrated with the sensor, the cavity is then assembled with the PMMA-film structure. Finally, inlet and outlet tubings are attached to the PMMA top and sealed using epoxy for injection and removal of liquid biosamples.

In the absence of channel, the recorded ψ and Δ spectrum of Au/MMA/Ge/Ag cavity in wavelength scan at 65° incidence angle is shown in Fig. 2.6a. As observed before, the cavity supports two modes in which longer wavelength mode that has resonance at 720 nm shows point-of-darkness and singular phase. The point-of-darkness and singular phase at 720 nm wavelength in angular scan is presented in Fig. 2.6c. In wavelength and angular scan, the measured ψ and Δ spectrum of PMMA microfluidic channel integrated cavity is shown in Fig. 2.6b, d, respectively. For

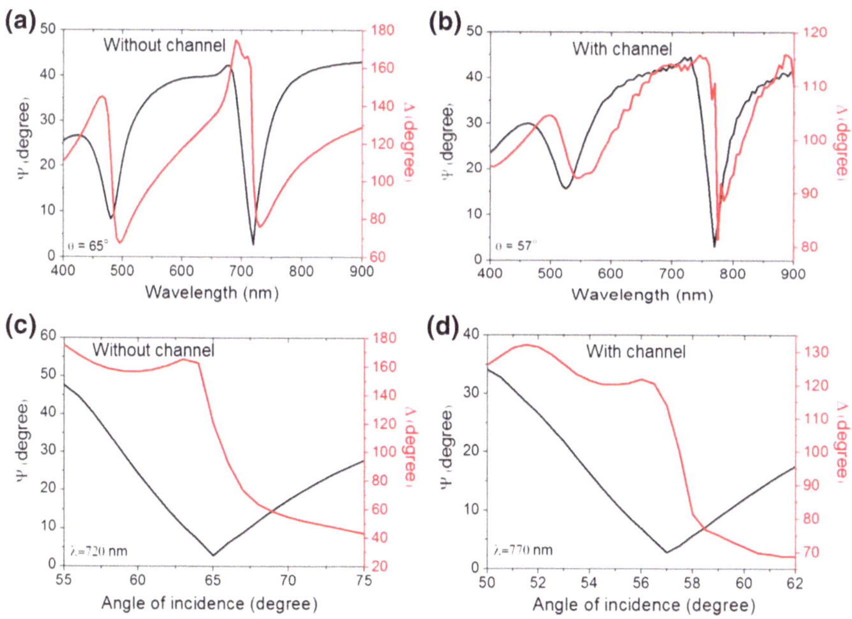

Fig. 2.6 Measured ellipsometry parameters (ψ and Δ) in wavelength scan **a** without channel at $\theta = 65°$ and **b** with channel at $\theta = 57°$. Measured ellipsometry parameters (ψ and Δ) in angular scan **c** without channel at $\lambda = 720$ nm and **d** with channel at $\lambda = 770$ nm. Reproduced with permission from Ref. [28]

microfluidics integrated cavity, the point-of-darkness and singular phase are realized at an incidence angle of 57°, as a result the spectral position of the mode is red shifted to 770 nm wavelength. This angle is considered to be the Brewster angle of the developed sensor device. Interestingly, the obtained ψ_{min} value and phase change at the point-of-darkness of the microfluidics integrated cavity is almost same as that of a bare cavity. In particular, near singular phase is achieved for the microfluidic channel integrated cavity, which is sufficient for biosensing applications.

2.6 Real-Time Sensing of Small Biomolecules

To demonstrate the real-time sensing in angular scan, we consider smaller scanning range close to the Brewster angle and use an angular step of 0.5° to capture the minute phase changes. In order to determine the bulk refractive index (RI) sensitivity of the sensor, initially, a standard sensor calibration test with known refractive index analytes is conducted. We injected different weight ratios (0.25–2.5% w/v) of aqueous solutions of glycerol with volume = 2.4 μl into the sensor device and analyzed the corresponding ψ and Δ shifts. The acquired ψ and Δ spectrum by injecting distilled (DI) water and 2% glycerol in DI water is shown in Fig. 2.7a. One can notice that a considerable Δ shift is recorded beyond the point-of-darkness (ψ_{min}) in comparison to ψ shift. This clearly indicates that a minute change in the refractive index of analyte causes a considerable change in phase close to the Brewster angle of the device. The marginal phase shift of glycerol solution calculated with respect to DI water ($\delta\Delta = \left|\Delta_{water} - \Delta_{glycerol}\right|$) with increase in refractive index is shown in Fig. 2.7b. It is clear that the phase shift varies linearly with increase in refractive index of analyte. To validate the calibration test data, we fitted the experimental phase shift data with numerical simulations (black line in Fig. 2.7b), which is based on a TMM model. As can be seen, the simulated phase shifts match very well with the experimental results. The parameters used to get best fit are: $\lambda = 770$ nm, $\theta = 55.005°$ and Au, MMA, Ge and Ag thickness are 20, 510, 10 and 75 nm, respectively. The obtained bulk refractive index sensitivity of the device is, $\delta\Delta/\delta n = 3333°$/RIU, which was calculated by using the Δ shift contrast between DI water and 0.5% glycerol. It should be noted that the obatined RI sensitivity is comparable with the angular sensitivity of a grating-coupled HMM sensor [29]. The RI sensitivity can be further enhanced by achieving extreme singular phase ($\psi_{min} < 1$) with microfluidics integrated cavity. A fivefold improvement in sensitivity is possible if the sensor works at the extreme singular phase [28].

At low analyte sample volumes, labeling techniques are mostly used to study the binding kinetics and detection of small molecules with molecular weight <500 Da. However, this approach is not preferable because the fluorescent tags used for conjugation may modify or prevent the functionality of the targeted molecules. Even though label-free optical techniques proposed so far provide more precise quantitative and kinetic measurements by monitoring the binding events of analytes in their normal forms, it is not possible to monitor the small molecule binding at low concentrations using the existing label-free optical techniques.

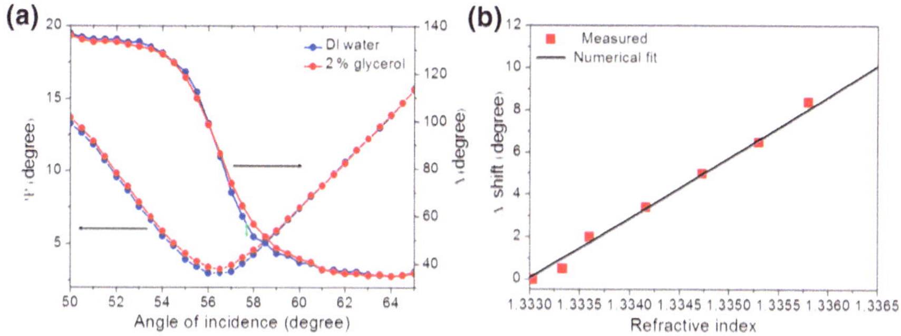

Fig. 2.7 Sensor calibration test by injecting different concentrations (0.25–2.5% w/v) of glycerol in DI water. **a** Measured ψ and Δ spectrum of DI water and 2% glycerol in DI water. **b** A plot of marginal phase shift ($\delta\Delta = |\Delta_{water} - \Delta_{glycerol}|$) with increase in refractive index of glycerol. The size of the data points (red squares) in **b** represents the error bar. Reproduced with permission from Ref. [28]

By using a microfluidics integrated cavity and a well-known streptavidin-biotin affinity model [21], we demonstrate the label-free detection and real-time binding of small biomolecules such as biotin in highly dilute solutions. We selected biotin as an analyte as it is a model system for small molecule compounds such as vitamins, cancer-specific proteins, hormones, and therapeutics. As a first step, the top Au surface of the cavity was functionalized with streptavidin proteins through a thiol-based surface chemistry in order to capture biotin. The sensor measures the refractive index changes caused by the capture of biotin proteins on the streptavidin sites. In particular, the performance of the sensor is monitored as a phase shift in the Δ spectrum by injecting different concentrations of biotin prepared in PBS (1 pM to 1 μM). In the sensing experiments, the spectrum of PBS is first obtained, and then the spectrum of different biotin concentrations is recorded after a reaction time of 30 min. Also note that PBS buffer is injected to remove the unbound and weakly attached biotin molecules before each injection of a new concentration of biotin solution. The acquired ψ and Δ spectrum for different concentrations of biotin is shown in Fig. 2.8a, b, respectively. In contrast to ψ shift, a large Δ shift is obtained with increasing biotin concentration, which is evident from the zoomed data shown in Fig. 2.8b. The phase change at the Brewster angle of the device by injecting 1 nM biotin is shown in Fig. 2.8c, where we plot the Δ spectrum as a function of incidence angle with a time step of 5 min. The increase in phase shift over time indicates the specific binding of biotin molecules on the streptavidin sites.

In Fig. 2.8d, e, we show the performance of sensor with different concentrations of biotin. The sensitivity of the phase shift to the number of adsorbed molecules on the sensor surface is analyzed. In this study, the saturation values of the phase shift for different biotin concentrations c is considered. In fact, the obtained phase shifts depend on the number of bound molecules $N(c)$ in the sensing area. Since it is not directly possible to measure the accurate value of $N(c)$, however, it is possible to estimate an upper bound $N_{\max}(c)$ based on the sensor parameters, such that the

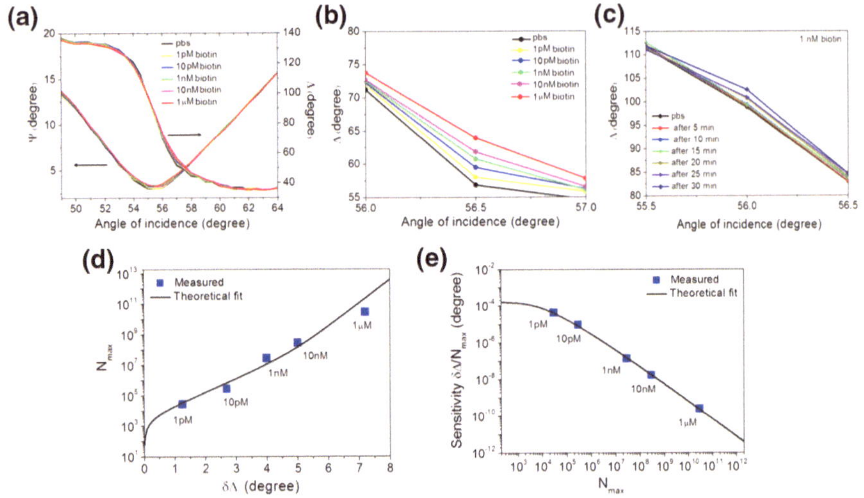

Fig. 2.8 **a** Measured ellipsometry parameters (ψ and Δ) of the sensor device for different concentrations (1 pM to 1 μM) of biotin in PBS. **b** Enlarged plots of Δ spectrum close to ψ_{min}, for different concentrations of biotin. **c** Experimental demonstration of real-time binding of biotin molecules on the sensor surface during the detection of 1 nM biotin in PBS. **d** The maximum number of biotin molecules (N_{max}) adsorbed in the illuminated sensor area versus the corresponding phase shift ($\delta\Delta$) and **e** the plot of the sensitivity $\delta\Delta/N_{\max}$ versus N_{max}. Each square (red) represents a different initial concentration of biotin, as indicated by the labels. The curve (black) is a best-fit function, described in the text. The size of the data points in **c** and **d** represents the error bar. Reproduced with permission from Ref. [28]

actual $N(c) \leq N_{\max}(c)$ for all concentrations [21]. By using a model described in [21], we obtained $N_{\max}(c) = 29c \times 10^{15}\ \mathrm{M}^{-1}$. The sensitivity of the phase shift ($\delta\Delta = |\Delta_{pbs} - \Delta_{biotin}|$) to the number of adsorbed molecules for the corresponding concentration of biotin is shown in Fig. 2.8d. It is clear that the experimental data points are accurately fitted (black curve) using a double-exponential fitting function, $N_{\max} = A_1(e^{\delta\Delta/\beta_1} - 1) + A_2(e^{\delta\Delta/\beta_2} - 1)$, where $A_1, A_2, \beta_1, \beta_2$, are fitting parameters and $\delta\Delta$ is the phase shift. The fitting parameters used to obtain the best fit are $A_1 = 3.1 \times 10^3$, $A_2 = 1.5$, $\beta_1 = 0.5°$, and $\beta_2 = 0.28°$. In Fig. 2.8e, we plot the sensitivity, $\delta\Delta/N_{\max}$ with $N_{\max}$, where the black line represents the fit obtained using the same model. One can see that the ratio, $\delta\Delta/N_{\max}$ decays roughly like $1/N_{\max}$ with increasing $N_{\max}$, which implies that sensitivity of the sensor is maximum at lower concentrations and it decreases with increasing concentration. Since the phase resolution of the VASE ellipsometer is <1°, it can be concluded that the detection limit that we achieved is 1 pM. It is important to note that the detection limit achieved for small molecules using the cavity system through phase-sensitive mechanism is comparable with that of nanopatterned HMM-based biosensors [21], however, where the sensing mechanism is based on spectral wavelength shift. The demonstrated label-free detection and real-time binding of small biomolecules using the proposed microfluidics integrated lithography-free nanophotonic cavities could be used to develop cost-effective non-invasive liquid biopsies for point-of-care clinical evaluation, early cancer screening and real-time diagnosis of diseases.

References

1. Aharonov Y, Bohm D (1959) Significance of electromagnetic potentials in the quantum theory. Phys Rev 115:485–491
2. Berry MV (1984) Quantal phase factors accompanying adiabatic changes. Proc R Soc Lond A Math Phys Sci 392:45–57
3. Kravets VG, Schedin F, Jalil R, Britnell L, Gorbachev RV, Ansell D, Thackray B, Novoselov KS, Geim AK, Kabashin AV, Grigorenko AN (2013) Singular phase nano-optics in plasmonic metamaterials for label-free single molecule detection. Nat Mater 12:304–309
4. Atwater HA, Polman A (2005) Plasmonics for improved photovoltaic devices. Nat Mater 9:205–213
5. Landy NI, Sajuyigbe S, Mock JJ, Smith DR, Padilla WJ (2008) Perfect metamaterial absorber. Phys Rev Lett 100:207402
6. Sreekanth KV, Alapan Y, Rashed AR, Gurkan UA, Strangi G (2016) A multiband perfect absorber based on hyperbolic metamaterials. Sci Rep 6:26272
7. Lin CH, Chern RL, Lin HY (2011) Polarization-independent broad-band nearly perfect absorbers in the visible regime. Opt Exp 19:415–424
8. Brewster D (1815) On the laws which regulate the polarization of light by reflection from transparent bodies. Philos Trans R Soc London 05:125–159
9. Raether H (1988) Surface plasmons on smooth and rough surfaces and on gratings. Springer, Berlin
10. Wan W, Chong Y, Ge L, Noh H, Stone AD, Cao H (2011) Time-reversed lasing and interferometric control of absorption. Science 331:889–892
11. Feng L, Xu YL, Fegadolli WS, Lu MH, Oliveira JEB, Almeida VR, Chen YF, Scherer A (2013) Experimental demonstration of a unidirectional reflectionless parity-time metamaterial at optical frequencies. Nat Mater 12:108–113
12. Hales TC (2007) The Jordon Curve theorem, formally and informally. Am Math Monthly 114:882–894
13. Malassis L, Masse P, Treguer-Delapierre M, Mornet S, Weisbecker P, Barois P, Simovski CR, Kravets VG, Grigorenko AN (2014) Topological darkness in self-assembled plasmonic metamaterials. Adv Mater 26:324–330
14. Aristov AI, Manousidaki M, Danilov A, Terzaki K, Fotakis C, Farsari M, Kabashin AV (2016) 3D plasmonic crystal metamaterials for ultra-sensitive biosensing. Sci Rep 6:25380
15. Song H, Zhang N, Duan J, Liu Z, Singer MH, Ji D, Cheney AR, Zeng X, Chen B, Jiang S, Gan Q (2017) Dispersion topological darkness at multiple wavelengths and polarization states. Adv Opt Mater 5:1700166
16. Svedendahl M, Verre R, Kall M (2014) Refractometric biosensing based on optical phase flips in sparse and short-range-ordered nanoplasmonic layers. Light Sci Appl 3:e220
17. Sreekanth KV, Sreejith S, Han S, Mishra A, Chen X, Sun H, Lim CT, Singh R (2018) Biosensing with the singular phase of an ultrathin metal-dielectric nanophotonic cavity. Nat Commun 8:36
18. Kats MA, Blanchard R, Genevet P, Capasso F (2013) Nanometre optical coatings based on strong interference effects in highly absorbing media. Nat Mater 12:20–24
19. Brolo AG (2012) Plasmonics for future biosensors. Nat Photon 6:709–713
20. Homola J, Yee SS, Gauglitz G (1999) Surface plasmon resonance sensors: review. Sens Actuators B 54:3–15
21. Sreekanth KV, Alapan Y, ElKabbash M, Ilker E, Hinczewski M, Gurkan UA, De Luca A, Strangi G (2016) Extreme sensitivity biosensing platform based on hyperbolic metamaterials. Nat Mater 15:621–627

22. Zijlstra P, Paulo PMR, Orrit M (2012) Optical detection of single non-absorbing molecules using the surface plasmon resonance of a gold nanorod. Nat Nanotech 7:379–382
23. Wu C, Adato R, Arju N, Yanikk AA, Altug H, Shvets G (2012) Fano-resonant asymmetric metamaterials for ultrasensitive spectroscopy and identification of molecular manolayers. Nat Mater 11:69–75
24. Konopsky VN, Alieva EV (2007) Photonic crystal surface waves for optical biosensors. Anal Chem 79:4729–4735
25. Vollmer F, Arnold S (2008) Whispering-gallery-mode biosensing: label-free detection down to single molecules. Nat Methods 5:591–596
26. Kabashin AV, Nikitin PI (1997) Interferometer based on a surface-plasmon resonance for sensor applications. Quant Electron 27:653–654
27. Wu SY, Ho HP, Law WC, Lin C, Kong SK (2004) Highly sensitive differential phase sensitive surface plasmon resonance biosensor based on the Mach-Zehnder configuration. Opt Lett 29:2378–2380
28. Sreekanth KV, Sreejith S, Alapan Y, Sitti M, Lim CT, Singh R (2019) Microfluidics integrated lithography-free nanophotonic biosensor for the detection of small molecules. Adv Opt Mater 7:1801313
29. Sreekanth KV, Alapan Y, ElKabbash M, Wen AM, Ilker E, Hinczewski M, Gurkan UA, Steinmetz NF, Strangi G (2016) Enhancing the angular sensitivity of plasmonic sensors using hyperbolic metamaterials. Adv Opt Mater 4:1767–1772

Chapter 3
Phase Change Material-Based Nanophotonic Cavities for Reconfigurable Photonic Device Applications

The discovery of the interesting optical properties of phase change materials (PCM) may open the door for photonic data storage again. These materials modify their optical constants which can be rapidly activated thermally, optically, or electronically [1]. In this chapter, we discuss the optical properties of phase change material such as Sb_2S_3 and $Ge_2Sb_2Te_5$ for visible and near infrared (NIR) photonics applications. In particular, we demonstrate PCM-based planar nanophotonic cavities for realizing tunable color filters, tunable wide-angle perfect absorption and tunable phase singularity, and Goos-Hänchen shift.

3.1 Phase Change Material-Tuned Photonics

The components of future active photonic devices could be tunable and reconfigurable. In particular, light strongly interacts with optical structures whose unit cell dimension similar to scale of the light wavelength, usually nanoscale features for light in the visible wavelength range. However, in the visible and NIR frequencies, the optical response of most of the nanostructures is fixed during the fabrication process. For photonics applications, phase change materials offer one way to tune the properties of optical structures. In phase change material-based active photonics, the refractive index of the material is tuned by switching the structural state from amorphous to crystalline. As an important advantage, PCM can be reversibly switched on a sub-nanosecond time scale billions of times, for example $Ge_2Sb_2Te_5$. In addition, no energy is needed to maintain the switched state once it is switched [2]. This feature allows the photonics devices to be programmed to have a specific response. The prototypical phase change material, $Ge_2Sb_2Te_5$ exhibits a substantial refractive index change when switched between the states from amorphous to crystalline [1]. In comparison to liquid crystals and mechanically tuned photonic devices, PCMs maintain their structural state and the energy is only required during the switching process, which is the key advantage. For optical and electrical data storage applications, phase change switching in tellurium-based materials has

Sreekanth K. V. et al., *New Directions in Thin Film Nanophotonics*, Progress in Optical Science and Photonics 6, https://doi.org/10.1007/978-981-13-8891-0_3

already been commercialized [3]. Moreover, thin film Fabry-Perot (F-P) cavities and plasmonic nanogratings based switchable reactive color display devices have been demonstrated using $Ge_2Sb_2Te_5$ [4–6]. However, particularly in the visible frequencies, $Ge_2Sb_2Te_5$ is not an ideal material to tune visible photonics devices due to the high absorption at above bandgap photon energies and the relatively small change in the real component of the refractive index [7]. The real and imaginary refractive indices of $Ge_2Sb_2Te_5$ in both amorphous and crystalline phases are shown in Fig. 3.1a. Note that ultra-thin film of $Ge_2Sb_2Te_5$ layers were used in switchable reactive color display based on Fabry-Perot cavities to minimize excessive absorption of the visible light [4]. However, it further restricts the phase transition induced optical path length change. Even though $Ge_2Sb_2Te_5$ displays small change in refractive index and higher absorption in the visible, the potential of $Ge_2Sb_2Te_5$-based active photonics is clear in the infrared, where the refractive index change is larger, and absorption is lower.

Thus, phase chase change materials with a low optical absorption in the visible spectrum are preferred over strongly absorbing materials, such as $Ge_2Sb_2Te_5$ for tunable visible photonics applications. It is clear from Fig. 3.1b that stibnite (Sb_2S_3) is a wide bandgap phase change material with a tunable absorption edge are highly suitable for active visible photonics [8]. The real permittivity values of Sb_2S_3 exhibit a substantial change at visible frequencies when the phase of the material is switched from amorphous to crystalline, while the imaginary permittivity is very small for wavelengths greater than 500 nm for both phases. It indicates that Sb_2S_3 is an appropriate low-loss and tunable dielectric material for the realization of active photonic devices at visible frequencies. The study of Sb_2S_3 was started in the 1990s for rewritable optical data storage [9]. However, Sb_2S_3 is not suitable for optical data storage because the Sb_2S_3 bandgap is larger than the photon energy of the diode lasers that were used to read from and write to compact discs. We find that Sb_2S_3

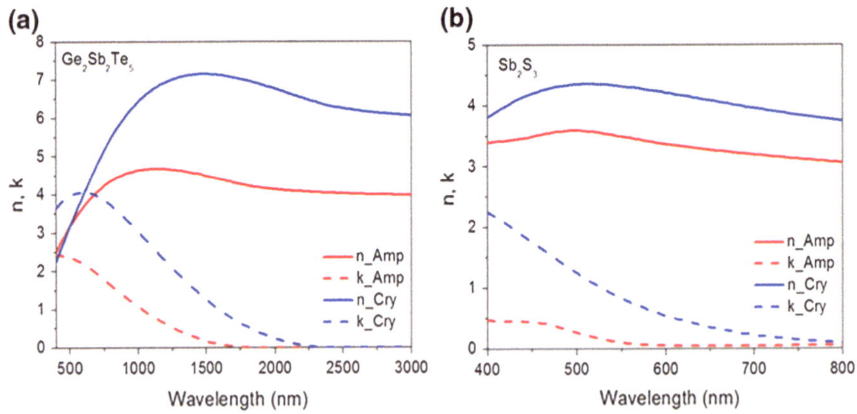

Fig. 3.1 Experimentally determined optical constants (n and k) of PCM thin films: **a** $Ge_2Sb_2Te_5$ and **b** Sb_2S_3

is an important reversibly tunable material for active visible photonics due to the following reasons [8]: (i) Sb_2S_3 is a wide band gap PCM that exhibits a large absorption edge shift at visible frequencies, (ii) Sb_2S_3 exhibits a large change to the real part of its refractive index in the visible spectrum during the structural phase transition, (iii) Both the amorphous and crystalline phases of Sb_2S_3 are stable at room temperature, which is important for non-volatile programming of optical devices, (iv) The phase transition temperatures of Sb_2S_3 are accessible by diode laser heating and (v) Sb_2S_3 has a crystallization activation energy of 2.0 eV, which is similar to that of $Ge_2Sb_2Te_5$ (2.3 eV).

3.2 Tunable Color Filters Based on Multilayer Stacks

As mentioned above, PCMs offer a way to tune the properties of optical structures in nanoseconds. To date, highly absorbing $Ge_2Sb_2Te_5$ has been used for phase change active photonics, which limits their application in the visible spectrum. As shown above, Sb_2S_3 is an underexplored phase change material with large refractive index contrast in the visible spectrum. In this section, we discuss that this tunable bandgap can be deliberately coupled to an optical resonator such that it shows tunable colors in the visible spectrum [8].

An $Al/Si_3N_4/Sb_2S_3/Si_3N_4/Al$ stacked structure was designed and fabricated to show the effect of using the Sb_2S_3 absorption edge change to tune at visible wavelengths. Figure 3.2a illustrates the fabricated device structure. This asymmetric (F-P) cavity structure was chosen to demonstrate the superior performance of Sb_2S_3, rather than a new type of photonics device. In the structure, the thickness of bottom and top Al is 100 and 3 nm, respectively. To prevent the interlayer diffusion between metal and chalcogenide layers, a diffusion barrier of Si_3N_4 was deposited between the Al and Sb_2S_3 layers. This is known to be a critical issue when metals are in direct contact with phase change chalcogenides. The reason for the top Al layer is to increase the Q-factor of the resonator. Note that Sb_2S_3 layer thickness is a variable parameter (varied from 11 to 67 nm) to shift the peak absorptance wavelength.

As the wavelength is scanned from 800 to 400 nm, Sb_2S_3 effectively switches from a low-loss to lossy dielectric (Fig. 3.1b). The mechanism of the color generation in the device is that the structural phase transition of Sb_2S_3 red-shifts the resonator's natural frequency, and this causes the strong reflected color change. This color change is clearly visible from the photographs of the sample that are displayed in Fig. 3.2c. The photographs and corresponding reflectance spectra of the sample with 24 nm thick Sb_2S_3 is in amorphous and crystalline states are shown in Fig. 3.2d, e, respectively. It is evident from Fig. 3.2e that the absorptance peak is red-shifted by 110 nm from 520 to 630 nm. It is also shown in Ref. [8] that as the thickness of the Sb_2S_3 layer increases, the peak absorptance wavelength is red-shifted across the visible spectrum from 425 to 710 nm. The simulation results reproduce the key spectral features that are observed in the experiment.

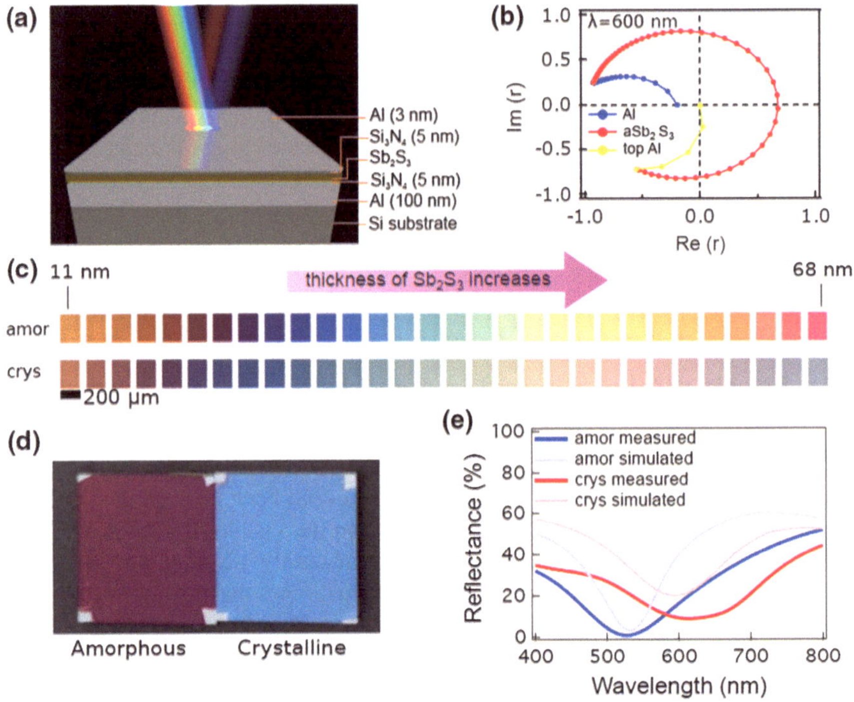

Fig. 3.2 **a** Schematic of the fabricated sample. **b** Calculated complex reflection coefficient at $\lambda =$ 600 nm for a cavity structure with thickness: 100 nm for Al layer, 51 nm for amorphous Sb_2S_3 layer, and 4 nm for top layer of Al. **c** Different colors achieved using the structure shown in **a** where the thickness of Sb_2S_3 is varied from 11 to 68 nm, with Sb_2S_3 in both amorphous and crystalline states. **d** Photograph of the color achieved with the 24 nm thick Sb_2S_3 layer in the amorphous and crystalline states. **e** Measured and simulated reflectance spectra of the structure shown in **a** with 24 nm thick Sb_2S_3 layer in the amorphous and crystalline states. Reproduced with permission from Ref. [8]

The red-shift observed in Fig. 3.2e is directly due to the intrinsic absorption edge of Sb_2S_3 narrowing from 2.0 to 1.7 eV. Specifically, the F-P cavity structure is designed to resonate at wavelengths close to the absorption-edge that amplifies the effect of the absorption edge switching on- and off- the resonant wavelength, which produces the large change in reflected color. However, as shown in Fig. 3.2e, the F-P cavity structure with crystalline Sb_2S_3 shows a broader resonance. This is because the crystalline Sb_2S_3 is slightly more absorbing than the amorphous state.

3.3 Tunable Perfect Absorption

Metamaterials and plasmonic nanostructures-based absorbers enabled controlled light absorption and achieved near-to-perfect absorption (<99%). In particular, metamaterials-based perfect absorbers are mainly focused on broadening the absorption band using different strategies, for example by overlapping multi-resonant impedance matching absorbers [10]. Introducing tunable materials into plasmonic and metamaterial designs enabled active devices with tunable absorption properties. By incorporating tunable materials such as graphene, liquid crystals and phase change materials (VO_2 and chalcogenides) into plasmonic systems different active plasmonic absorbers have been demonstrated [11–14]. The external stimulus such as thermal, electrical and optical create tunability in such systems. Since graphene, microelectromechanical systems (MEMS) and liquid crystal-based absorbers are usually tuned via electrically, however, as an advantage, phase change material-based systems can be tuned by electrically, thermally and optically. In addition, the active absorbers demonstrated so far particularly in the visible spectrum regime are lithographically intense, which restricts the scalability of the samples. Thus, the development of practical active photonic devices in the visible spectrum regime is currently facing many technological challenges. More importantly, realizing perfect light absorption in the visible frequencies using PCM is crucial as higher frequencies provide larger modulation bandwidth.

For realizing perfect absorption over a broad spectral band from visible to terahertz, the metal-dielectric stacks-based F-P cavity systems have attracted much interest in recent years since they possess a lithography-free design that is cost-effective and scalable. In addition, the multilayer thin-film stacks containing phase change materials such as VO_2 and $Ge_2Sb_2Te_5$ (GST) have been demonstrated for tunable perfect absorption [15–17]. However, the GST-based systems provide near-perfect absorption at visible and near infrared frequencies, whereas the VO_2-based systems provide perfect absorption at infrared frequencies only. The drastic reduction in resonant absorption after switching the GST phase from amorphous to crystalline state is one of the limitations of GST-based perfect absorbers [16].

Here, we propose an active nanophotonic cavity for tunable perfect absorption in the visible spectral band [18]. As shown in Fig. 3.3a, the F-P cavity consists of an 18 nm top and 50 nm bottom Ag layers forming the cavity mirrors and a thick lossless MMA layer is the cavity medium. An ultra-thin layer of PCM such as $Ge_2Sb_2Te_5$ with thickness 8 nm is introduced between MMA and the bottom Ag layers. GST layer is grown by RF sputtering of $Ge_2Sb_2Te_5$ alloy target at a deposition rate of 0.2 Å/s and a base pressure of $<5 \times 10^{-6}$ mbar. We selected GST as the phase change material because the cavity layer is a polymer. Note that the crystallization temperature of $Ge_2Sb_2Te_5$ and Sb_2S_3 is 160 and 280 °C, respectively. However, Sb_2S_3 can be used as an ultrathin absorbing layer by replacing MMA layer with SiO_2, so that higher tunability can be achieved in the visible frequencies.

By using a spectroscopic ellipsometer, the reflectance as a function of wavelength (400–1000 nm) are acquired for different angles of incidence (20°–80°).

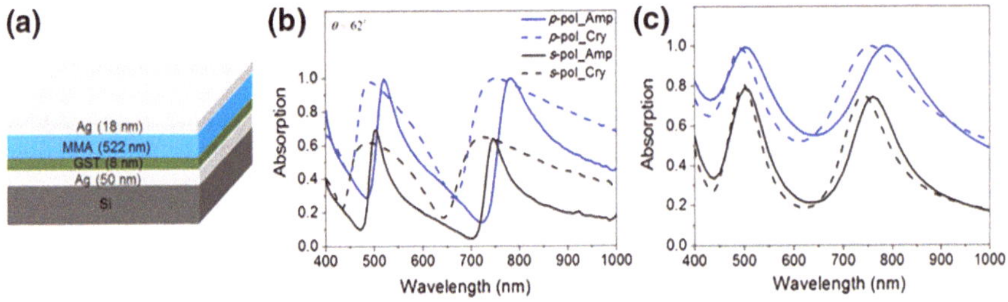

Fig. 3.3 **a** Schematic representation of the proposed nanophotonic cavity. The *p*- and *s*-polarized absorption spectrum of amorphous and crystalline stack at 62° **b** measured and **c** simulated. *Amp:* GST is in amorphous phase, *Cry:* GST is in crystalline phase. Reproduced with permission from Ref. [18]

Then, the absorptance (A) are obtained from the reflectance spectra (R), that is $A = 1 - R - T$, where T is the transmission, with $T = 0$. In Fig. 3.3b, we plot the as obtained absorptance of the PCM-based cavity when GST is in both amorphous and crystalline phase. For both *p*- and *s*-polarization, the F-P cavity supports two cavity modes. As can be seen, the maximum absorptance is obtained for *p*-polarization at the Brewster angle. For *as deposited* stack (GST is in amorphous) at 62° angle of incidence, around unity absorptance is realized, which is 99.7 and 99.8% at 520 and 782 nm, respectively. In order to switch the phase of the GST in the stack to crystalline, we baked the *as deposited* stack at 160 °C for 60 min and the measured absorptance is shown as dotted curves in Fig. 3.3b. A slightly decreased absorptance with a blue spectral shift and broadening of the mode is obtained for the crystalline GST stack at 62° angle of incidence. The recorded absorptance at 500 and 750 nm is 97.7 and 99.7%, respectively. However, in fact, crystalline GST stack provides slightly increased absorptance at the Brewster angle as the maximum absorptance is obtained at 736 nm and 66° angle of incidence, which is 99.97%. More importantly, the proposed tunable cavity shows perfect absorption in the visible wavelengths for both phases of GST, however, the existing thin-film based-tunable absorbers do not display these features. The experimental results are then corroborated with a TMM-based numerical simulation model. The simulated results are shown in Fig. 3.3c. As can be seen, the simulation results precisely reproduce the significant spectral features such as the blue-shifted absorptance maximum upon crystallization and the increased absorptance for *p*-polarization as compared to *s*-polarization. The tunable perfect absorption is highly sensitive to changes in the optical properties of GST which are activated by switching GST phase from amorphous to crystalline. In addition, the absorption resonance gets enhanced with increasing the complex refractive index of GST layer. That is why the stack with crystalline GST layer provided a slightly higher absorptance at the Brewster angle compared to the stack with amorphous GST.

To experimentally demonstrate the tunable wide-angle perfect absorption and the influence of GST layer thickness on tunable perfect absorption, we measured the *p*-polarized absorptance spectra as a function of the incidence angle (20°–80°) for different thicknesses of GST layer (8, 20 and 30 nm), which is shown in Fig. 3.4. For the three considered thicknesses, the perfect absorption bands broaden and blue shift when the GST layer in the stack is switched from amorphous crystalline state. It is important to notice that all angle and multiband tunable perfect absorption is only observed for the stack with 8 nm thick GST layer. It is due to the fact that a thick GST layer alters the conditions for critical light coupling to the resonator, as a result the light trapped inside the cavity decreases. We further consider longer wavelength absorption band (above 700 nm) because this band only show perfect absorption with increasing GST layer thickness. For the stack with amorphous GST layer, all angle perfect absorptance feature deviate with increasing GST layer thickness (Fig. 3.4a, d, g). However, all angle perfect absorptance is achieved for the stack with crystalline GST layer for the three thicknesses of GST (Fig. 3.4b, e, h). Moreover, broadening of the absorption band is visible in the case of crystalline GST stack with increasing GST layer thickness. As we obtained maximum absorptance for crystalline GST stack, the measured peak absorptance is 99.97, 99.8 and 99.6% for 8, 20 and 30 nm thick GST layer stack, respectively. The decrease of absorptance with increasing GST layer thickness evidences the losing of critical coupling conditions. We further calculated the percentage absorptance change when the GST structural phase is switched from amorphous to crystalline, which is $\Delta A(\%) = \frac{A_{Cry}-A_{Amp}}{A_{Amp}} \times 100$. The absorptance change for 8, 20 and 30 nm thick GST layer is shown in Fig. 3.4c, f, i, respectively. As can be seen, maximum percentage absorptance change is obtained for 8 nm thick GST layer, which is around eightfold compared to 30 nm thick GST layer. It indicates that the proposed nanophotonic cavity with an ultrathin GST (<10 nm) layer is required to achieve efficient tunable perfect absorption in the visible spectral band.

3.4 Tunable Singular Phase at the Point-of-Darkness

An important feature of the GST-based stack is that it exhibits tunable singular phase behavior at the point-of-darkness by switching the structural phase of GST layer from amorphous to crystalline. As shown above, the maximum p-polarized perfect absorptance for amorphous and crystalline GST stack is obtained at the Brewster angle of 62° and 66°, respectively, therefore an abrupt phase change is expected at those incidence angles. In Fig. 3.5a, we show the measured ψ and Δ spectrum of the amorphous and crystalline GST stack at 62° and 66°, respectively. As can be seen, the singular phase behavior at the point-of-darkness is obtained at 782 and 736 nm for amorphous and crystalline GST stack, respectively. At the point-of-darkness, the recorded ψ and phase change are 1.05°, 40° and 0.82°, 89° for amorphous and crystalline GST stack, respectively. It shows that the abrupt phase change increases with decreasing ψ_{min} value as higher phase change is obtained for crystalline GST stack compared to amorphous GST stack. Besides, the broadening of ψ spectra is observed for the crystalline GST stack, which is consistent with absorptance spectra

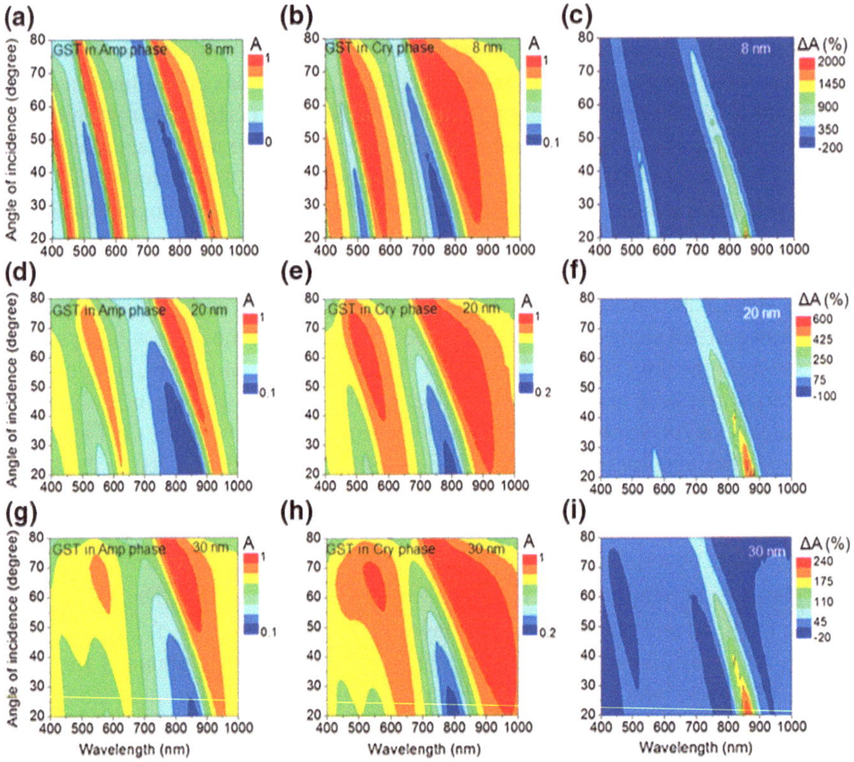

Fig. 3.4 Measured 2D map of absorptance and percentage absorptance for different thicknesses of GST layer. For stack with GST is in amorphous phase **a** 8 nm GST, **d** 20 nm GST, and **g** 30 nm GST. For stack with GST is in crystalline phase **b** 8 nm GST, **e** 20 nm GST, and **h** 30 nm GST. Percentage absorption change of stack with **c** 8 nm GST, **f** 20 nm GST, and **i** 30 nm GST. Reproduced with permission from Ref. [18]

of the crystalline GST stack. As mentioned above, the perfect absorption is even possible without the GST layer, however, the tunable singular phase behavior is only possible with an ultra-thin GST film in the stack. In short, the point-of-darkness and thus the abrupt phase change can be tuned by switching the structural phase of the GST between amorphous and crystalline states.

The simulated pair of ellipsometry parameters (ψ and Δ) using a TMM model is shown in Fig. 3.5b. One can see that the simulated and measured spectra match very well with each other. In the simulation, the thickness of the individual layers of the stack are slightly varied to obtain a best fit. We further used the TMM simulation model to show the confined intensity distribution along the planar multilayer stack. For both phases of GST, the simulated intensity distribution of the two cavity modes are shown in Fig. 3.6. Figure 3.6a, b, respectively represents the field map of the mode at 520 nm and mode at 782 nm for amorphous GST stack at 62° angle of incidence, whereas, the corresponding field obtained at 66° for crystalline GST stack is shown

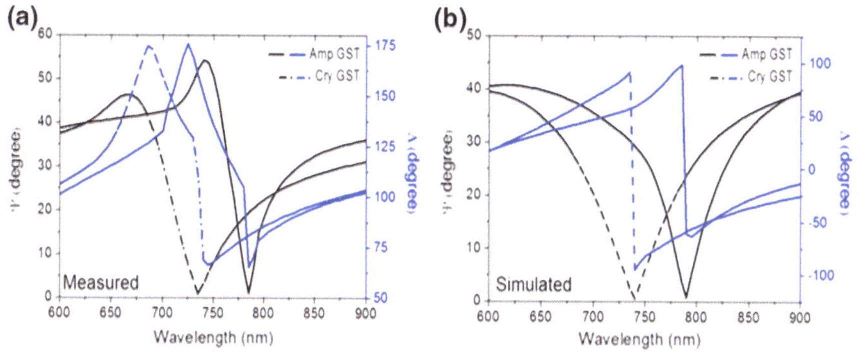

Fig. 3.5 Demonstration of tunable singular phase at the point-of-darkness. Pair of ellipsometry parameters as a function of wavelength for both phases of GST in the stack **a** measured and **b** simulated

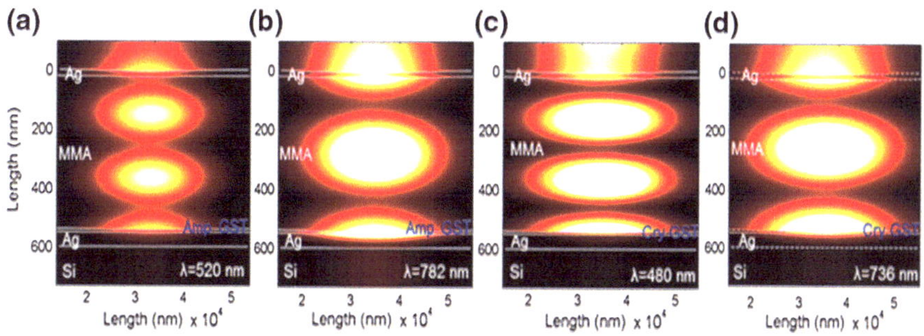

Fig. 3.6 Simulated intensity field distribution along the PCM-based cavity using TMM. For amorphous GST stack at 62°: **a** mode at 520 nm and **d** mode at 782 nm. For crystalline GST stack 66°: **c** mode at 480 nm and **d** mode at 736 nm. Reproduced with permission from Ref. [18]

in Fig. 3.6c for the mode at 480 nm and in Fig. 3.6d for the mode at 736 nm. It is clear that the field intensity is strongly confined in the ultrathin GST layer for the two modes and both phases of GST, however, the confined field intensity in the case of longer wavelength mode is much higher compared to the shorter wavelength mode. Also note that the crystalline GST stack shows reduced field confinement with respect to amorphous GST stack, which is due to the broadening of absorption spectra after crystallization of the GST layer. The obtained numerical simulation results further support the important role of GST in the planar multilayer stack for tunable perfect absorption.

3.5 Enhanced and Tunable Goos-Hanchen Shift at the Point-of-Darkness

In geometrical optics, Goos-Hänchen (G-H) shift is a well-known phenomenon, which describes the lateral displacement of the reflected beam from the interface of two media when the angles of incidence are close to the critical angle [19]. In particular, this lateral beam displacement usually occurs when the light is totally reflected from an interface of two media and it is proportional to the penetration depth of the light field. The G-H shift is also observed when the light is partially reflected or transmitted from an interface of two media. However, the direct observation of G-H shift is very challenging as the shift is usually of the order of the light wavelength. Therefore, different approaches have been demonstrated to enhance the G-H shifts [20]. The researchers have proposed two important models to interpret the appearance of G-H shift, which are stationary phase approach and energy propagation approximation. A general formula for calculating G-H shift based on stationary phase approach is, $S = -\frac{1}{k}\frac{\partial\phi}{\partial\theta}$, where k is the wavevector of the incident medium and ϕ is the phase of the reflection coefficient. Here, the G-H shift is related to the change in the complex reflection coefficient phase. It implies that the phase change of light at the resonance angle determines the magnitude of the G-H shifts.

In recent years, the G-H shift enhancement has been theoretically and experimentally demonstrated using different optical systems such as structural resonances, photonic crystals, absorptive materials, metals, negative refractive index media and hyperbolic metamaterials [21–25]. The reported G-H shifts in the case of well-known interfaces such as 'in the vicinity total internal reflection angle', 'beyond the Brewster angle' and 'at the edge of the photonic band gap', are of the order tenfold of the operating wavelength [26–28]. This is because these interfaces usually show small phase derivative at the minimum reflectance angle (or resonant angle) of their broad spectrum. However, by exciting the surface plasmons on metal surfaces [29] and the surface electromagnetic waves (SEWs) of photonic crystals [30], a significant enhancement in G-H shift has been experimentally demonstrated. This enhanced G-H shift is due to the large phase derivative at the resonant angle as surface waves provide high quality factor modes. Though, in order to excite the surface waves of metals and photonic crystals, the coupling techniques such as prism or grating is required.

Here, we show that enhanced and tunable G-H shift is possible at the Brewster angle of the proposed nanophotonic cavity [31]. In order to experimentally demonstrate enhanced G-H shift, we designed a nanophotonic cavity that shows point-of-darkness at 633 nm wavelength. The fabricated Ge-based four layered cavity is shown in the inset of Fig. 3.7a, where the thickness of top Ag, MMA, Ge and bottom Ag layer is 20, 400, 8 and 80 nm, respectively. The measured ellipsometry parameters (ψ and Δ) of the cavity as a function of wavelength and angle of incidence are shown in Fig. 3.7a, b, respectively. As expected, the cavity shows point-of-darkness ($\psi < 1$) at 633 nm wavelength and 65.5° angle of incidence. As a result, abrupt phase change is obtained at a Brewster angle of 65.5° and thus enhanced G-H shift is expected.

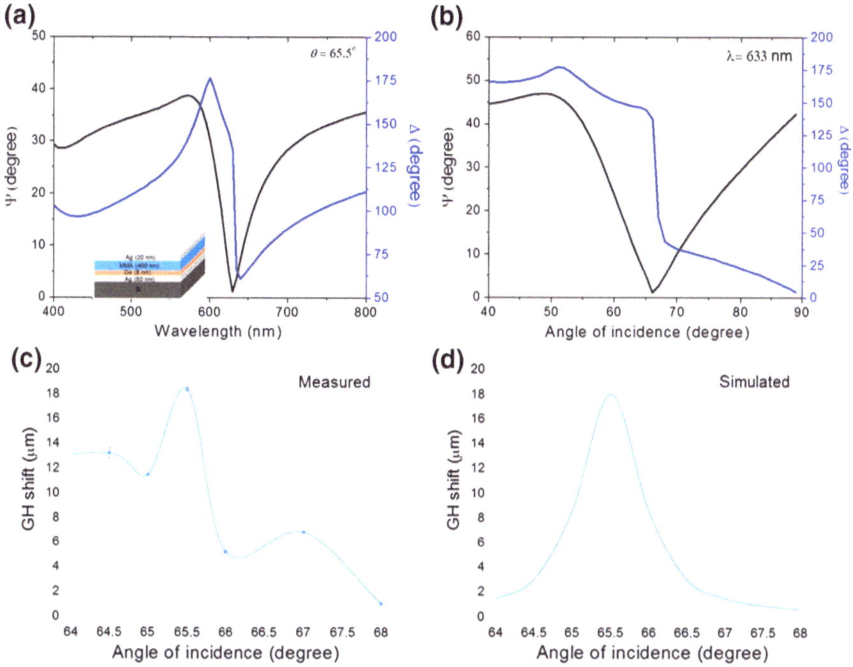

Fig. 3.7 Measured ψ and Δ spectrum **a** as a function of wavelength at 65.5° angle of incidence, **b** as a function of incidence angle at 633 nm wavelength. G-H shift as a function of incidence angle at 632.8 nm wavelength **c** measured and **d** simulated. Data points in **c** represent the error bar. Reproduced with permission from Ref. [31]

A home-built differential phase-sensitive optical setup is used to experimentally demonstrate the G-H shift [31]. In the experiment, we recorded the differential G-H shift between *p*- and *s*-polarized light since *p*-polarized light only shows phase change at the Brewster angle. However, *s*-polarized light acts as a reference signal as it does not show any phase change. A 632.8 nm He–Ne laser with a power of 4 mW is used as the excitation source. By using a high-resolution lateral position sensor (Thorlabs, PDP90A), the positions of the reflected *p*- and *s*-polarized light as a function of incidence angle are recorded and a data acquisition card is used for further processing. A Matlab-based numerical code is used to obtain the differential G-H shift. Even though the position detector record the exact position of the light spot for *p*-polarized light, i.e. P (x_p, y_p) and for *s*-polarized light, i.e. S (x_s, y_s), we only considered the x-axis value of the spot position to calculate the differential G-H shift, $\Delta GH = |x_p - x_s|$. This is because the incident sample surface is kept parallel to the x-axis of the detector. A total of ~1500 points in 5 min of the signal recorded for each angle of incidence, where the sampling frequency of the detector is 50 Hz. As a first step, we calculated the standard deviation of these 1500 points of the data and then we filtered the original data by using the 2 times of the standard deviation. We then calculated the average value of both polarizations from the filtered data, which are the targeted results of both polarizations. The actual differential G-H shift is finally obtained from the absolute difference between these two averages. The observed noise level in the measurement is of the order of 0.25–0.3 μm.

As recorded G-H shift as a function of incidence angle is shown in Fig. 3.7c. Note that we acquired date from different positions of the sample to ensure reproducibility of the results. Since the measurable G-H shift is only expected close to the Brewster angle, we performed the G-H shift measurement in the angle range from 64° to 68°. As can be seen, maximum G-H is recorded at the Brewster angle (65.5°), which is nearly 30-fold of the excitation wavelength at 632.8 nm. It is important to notice that this G-H enhancement is about three times higher than that realized at the Brewster angle of a planar metal film [27].

A TMM-based numerical model demonstrating the G-H shift of the proposed experimental system is shown in Fig. 3.7d. In the calculation, the G-H shift is obtained by numerical differentiation of the formula, $S = -\frac{1}{k}\frac{d\phi}{d\theta}$ with θ is the angle of incidence, and ϕ is the phase angle of reflection coefficient ($\tilde{r}$). In fact, the simulation results show the expected G-H shift curve in angular scan. However, the experimental results only accurately reproduce the key results that the maximum G-H shift is obtained at the Brewster angle. The observed variation in the experimental G-H shift curve close to the Brewster angle is attributed to the limitation of the measurement apparatus. This is because the incident beam position on the cavity surface should not be the same at higher angles of incidence. As a result, random phase change occurs due to the slight displacement of the beam position, which in turns effect the absolute difference between the two average G-H shifts.

It is important to note that the G-H shift can make tunable by replacing Ge layer with a PCM layer. The numerical results demonstrating the tunable G-H shift using GST layer is shown in Fig. 3.8a. The G-H shift tunability in the visible spectral band can be further improved by using Sb_2S_3 because Sb_2S_3 provides higher refractive index contrast between amorphous and crystalline phase. To demonstrate the reversible electrical switching, we modeled the device design to electrical heat the PCM-based cavity. As shown in Fig. 3.8b, the design is based on a planar phase change line cell. The operational principle of the device is based on the electro-thermal effect. In a bowtie structure, the current density is highest at the narrowest point in the bowtie's taper. Consequently, the hottest point in the structure is also at its narrowest point. This has the advantage that only a small volume of material reaches a high temperature. And therefore, the amount of thermal energy stored in the structure is relatively low. In particular, the wider areas of the bowtie structure, which are at a low temperature, can act as a heat sink to remove heat from the phase change material quickly; this is necessary for amorphization. We have shown by modelling that even with modest voltages it is possible to heat phase change materials above their melting temperatures using a bowtie structure. Figure 3.8b shows a design that we have simulated. The device comprises of two tapered gold electrodes, which are connected through a gold line-cell. The gold rod is capped with a phase change material, $Ge_2Sb_2Te_5$ layer and this is protected by SiO_2 cavity layer as a cap. Our initial simulations, where we numerically solve the time-dependent Fourier equation in 3D

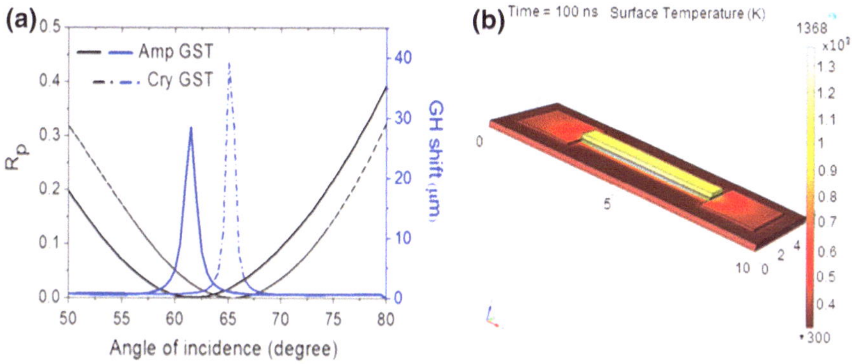

Fig. 3.8 **a** Numerical demonstration of tunable G-H shift and **b** Simulation for an electrically switchable PCM–based cavity

using a finite-element approach (COMSOL) that couples thermal and electric joule heating physics, shows that the $Ge_2Sb_2Te_5$ material reaches 1300 K in 100 ns, which is sufficient to induce the phase change. We also expect similar designs can be used for heating Sb_2S_3-based cavity.

References

1. Wuttig M, Bhaskaran H, Taubner T (2017) Phase-change materials for non-volatile photonic applications. Nat Photon 11:465–476
2. Qian W, Rogers ETF, Gholipour B, Yuan G, Teng J, Zheludev NI (2016) Optically reconfigurable metasurfaces and photonic devices based on phase change materials. Nat Photon 10:60
3. Zhu Z, Evans PG, Haglund RF Jr, Valentine JG (2017) Dynamically reconfigurable metadevice employing nanostructured phase-change material. Nano Lett 17:4881–4885
4. Hosseini P, Wright CD, Bhaskaran H (2014) An optoelectronic framework enabled by low-dimensional phase change films. Nature 511:206
5. Yoo S, Gwon T, Eom T, Kim S, Hwang CS (2016) Multicolor changeable optical coating by adopting multiple layers of ultrathin phase change material film. ACS Photonics 3:1265
6. Dong W, Qiu Y, Yang J, Simpson RE, Cao T (2016) Wideband absorbers in the visible with ultrathin plasmonic-phase change material nanogratings. J Phys Chem C 120:12713
7. Dong W, Qiu Y, Zhou X, Banas K, Breese MBH, Cao T, Simpson RE (2018) Tunable mid-infrred phase change metasurface. Adv Opt Mater 6:1701346
8. Dong W, Liu H, Behera JK, Lu L, Ng RJH, Sreekanth KV, Zhou X, Yang JKW, Simpson RE (2019) Wide bandgap phase change material tuned visible photonics. Adv Funct Mater 29:1806181
9. Arun P, Vedeshwar A (1997) Effect of heat treatment on the optical properties of amorphous Sb_2S_3 film: the possibility of optical storage. J Non-Cryst Solids 220:63
10. Aydin K, Ferry VE, Briggs RM, Atwater HA (2011) Broadband polarization-independent resonant light absorption using ultrathin plasmonic super absorbers. Nat Commun 2:517
11. Fang Z, Wang Y, Schlather AE, Liu Z, Ajayan PM, García Javier, de Abajo F, Nordlande P, Zhu X, Halas NJ (2014) Active tunable absorption enhancement with graphene nanodisk arrays. Nano Lett 14:299–304

12. Kats MA, Blanchard R, Genevet P, Yang Z, Qazilbash MM, Basov DN, Ramanathan S, Capasso F (2013) Thermal tuning of mid-infrared plasmonic antenna arrays using a phase change material. Opt Lett 38:368–370
13. Pradhan JK, Ramakrishna SA, Rajeswaran B, Umarji AM, Achanta VG, Agarwal AK, Ghosh A (2017) High contrast switchability of VO_2 based metamaterial absorbers with ITO ground plane. Opt Exp 25:9116
14. Su Z, Yin J, Zhao X (2015) Soft and broadband infrared metamaterial absorber based on gold nanorod/liquid crystal hybrid with tunable total absorption. Sci Rep 5:16698
15. Rensberg J, Zhou Y, Richter S, Wan C, Zhang S, Schöppe P, Schmidt-Grund R, Ramanathan S, Capasso F, Kats MA, Ronning C (2017) Epsilon-near-zero substrate engineering for ultrathin-film perfect absorbers. Phys Rev Appl 8:014009
16. Kats MA, Sharma D, Lin J, Genevet P, Blanchard R, Yang Z, Qazilbash MM, Basov DN, Ramanathan S, Capasso F (2012) Ultra-thin perfect absorber employing a tunable phase change material. Appl Phys Lett 101:221101
17. Mkhitaryan VK, Ghosh DS, Rudé M, Canet-Ferrer J, Maniyara RA, Gopalan KK, Pruneri V (2017) Tunable complete optical absorption in multilayer structures including $Ge_2Sb_2Te_5$ without lithographic patterns. Adv Opt Mater 5:1600452
18. Sreekanth KV, Han S, Singh R (2018) $Ge_2Sb_2Te_5$-based tunable perfect absorber cavity with phase singularity at visible frequencies. Adv Mater 30:706696
19. Goos F, Hanchen H (1947) Ein neuer und fundamentaler Versuch zur Totalreflexion. Ann Phys 436:333–346
20. Bliokh KY, Aiello A (2013) Goos-Hänchen and Imbert-Fedorov beam shifts: an overview. J Opt 15:014001
21. Yang R, Zhu W, Li J (2014) Giant positive and negative Goos-Hänchen shift on dielectric gratings caused by guided mode resonance. Opt Exp 22:2043–2050
22. Pfleghaar E, Marseille A, Weis A (1993) Quantitative investigation of the effect of resonant absorbers on the Goos-Hänchen shift. Phys Rev Lett 70:2281
23. Berman PR (2002) Goos-Hänchen shift in negatively refractive media. Phys Rev E 66:067603
24. Luo L, Tang T (2016) Goos-Hänchen effect in Kretschmann configuration with hyperbolic metamaterials. Superlattices Microstruct 94:85–92
25. Liu X, Cao Z, Zhu P, Shen Q, Liu X (2006) Large positive and negative lateral optical beam shift in prism-waveguide coupling system. Phys Rev E 73:056617
26. Schwefel HGL, Köhler W, Lu ZH, Fan J, Wang LJ (2008) Direct experimental observation of the single reflection optical Goos-Hänchen shift. Opt Lett 33:794
27. Merano M, Aiello A, t' Hooft GW, van Exter MP, Eliel ER, Woerdman JP (2007) Observation of Goos-Hänchen shifts in metallic reflection. Opt Exp 15:15928
28. Felbacq D (2003) Goos-Hänchen effect in the gaps of photonic crystals. Opt Lett 28:1633
29. Yin X, Hesselink L (2006) Goos-Hänchen shift surface plasmon resonance sensor. Appl Phys Lett 89:261108
30. Wan Y, Zheng Z, Kong W, Zhao X, Liu Y, Bian Y, Liu J (2012) Nearly three orders of magnitude enhancement of Goos-Hanchen shift by exciting Bloch surface wave. Opt Exp 20:8998–9003
31. Sreekanth KV, Ouyang Q, Han S, Yong KT, Singh R (2018) Giant enhancement in Goos-Hänchen shift at the singular phase of a nanophotonic cavity. Appl Phys Lett 112:161109

Chapter 4
Sub-wavelength Nanopatterning Using Thin Metal Films

Over the decade, photolithography has played a vital role in almost every aspects of modern technology. Developments in this field have allowed researchers to improve the resolution of the conventional photolithographic techniques that is restricted by the optical diffraction limit [1]. However, the implementation of recently developed high resolution lithography techniques beyond the diffraction limit are challenging because of their significant design and layout issues. In this chapter, single-exposure multiple-beams interference lithography generated by thin metal films are demonstrated, which overcome the issues associted with the optical diffraction limit. The proposed two-beam and four-beam surface plasmon interference approaches are carried out theoretically and numerically, and then verified experimentally.

4.1 Laser Interference Lithography

In nanotechnology research, interference lithography techniques play a major role in fabricating wide variety of complex structures. It is a maskless, powerful and cost-effective optical lithography technique to realize periodic structures over large-area (up to ~cm^2) at subwavelength scale resolution. The basic principle of laser interference lithography (LIL) is the interference of two or more coherent light beams from a laser source to form a horizontal standing wave pattern in the far-field, which can be recorded either in a photoresist or directly in the material of interest [2]. The recorded interference pattern period on the photoresist layer is given by $\Lambda = \lambda/2sin\theta$, where λ is the incident illumination laser wavelength and θ is the interfering angle with respect to the normal.

Laser interference lithography can be classified into two, named by wavefront splitting [3] and amplitude splitting [4]. Amplitude splitting-based interference technique is the most commonly used interference lithography to fabricate large-area periodic structures. The single exposure two beams interference lithography creates one-dimensional grating lines, whereas a second exposure with sample rotation of 90° and 60° is required to pattern hole/dot arrays with square and hexagonal

Sreekanth K. V. et al., *New Directions in Thin Film Nanophotonics*, Progress in Optical Science and Photonics 6, https://doi.org/10.1007/978-981-13-8891-0_4

(triangle) symmetry, respectively. In addition, multiple beams single exposure interference lithography can be used to pattern structures with square and hexagonal lattice symmetry. A multiple beams interference lithography enables the fabrication of nanoscale periodic features with two or more dimensions, which can find potential applications in photonic crystals, field emitter arrays, and biosensors. It has been shown that there is a tremendous significance in multiple beams interference as a fabrication technique for realizing advanced opto-electronic devices. The major advantages of multiple beams interference lithography are: (i) the multiple exposures are not required, and (ii) it eliminates the associated rotation of sample in between the exposures.

In contrast to amplitude splitting techniques, wavefront splitting techniques are simple to use and easy to control. Commonly used wavefront splitting techniques are based on the biprism [5] and the Lloyd's mirror [6] configurations. The mass production of nanoscale periodic structures is possible by combining these techniques with fast alignment and easily adjustable pattern size. In particular, biprism-based wavefront splitting technique is a simple and efficient laser interference lithography tool for fabricating periodic structures. In this case, both the interference pattern period and the resolution of the structures can be further reduced by using large angle and high refractive index biprisms.

Even though laser interference lithography produces large-area patterns, it is not possible to pattern sub-wavelength or high-resolution structures due to the optical diffraction limit, particularly using visible and near-ultraviolet (UV) wavelength sources. More specifically, there are several critical challenges to be faced to fabricate periodic structures with sub-100 nm critical resolution. To achieve this goal, possible options are the use of: (i) near-field lithography techniques and (ii) shorter wavelength light sources.

4.2 Evanescent Wave Interference Lithography

In recent years, the introduction of various near-field lithography techniques have received much attention in the resolution enhancement purpose. One such cost-effective near-field technique is the evanescent wave interference lithography (EWIL). There are different approaches to generate evanescent field such as by using total internal reflection (TIR) in an incident medium of high refractive index and using sub-wavelength features. As a first demonstration of EWIL, solid immersion lens (SIL) technique has been used as it provides high transmission or reflection efficiency and fast imaging possibility [7]. In the case of SIL, high refractive index solid immersion lens is combined with conventional far-field optical system. Note that the critical dimension of the image is inversely proportional to refractive index of the SIL.

There are two main methods to apply the solid immersion technique to optical lithography to improve the critical resolution of the structures. The first one is based on the hemispherical solid immersion lens [8]. The focusing of beam inside a hemispherical SIL is possible when the incident light illuminates on the top hemispherical surface, from a range of low angle to high angle. Two parts of the incident energy on the bottom planar surface of hemispherical SIL is possible. The first part of the incident energy propagates out of the SIL that is a homogeneous cone of energy formed from the portion of the beam focused at low incidence angles. The second part of incident energy is around the homogeneous cone of energy. More importantly, this is an annular ring of evanescent waves (EWs) arises from the total internal reflection of the portion of the beam at high incidence angles. The contribution of both waves, one that propagates out of the SIL and other that propagates on the bottom surface of the SIL, are responsible for the resultant high-resolution spot at the imaging plane. Also note that the sharpest features of the focused spot define the EWs. However, one of the limitations of this technique is that the complex optical system such as micro lens array is required for the realization of parallel SIL writing system.

Another promising method of solid immersion technique is by exploiting the interference of the evanescent waves [9]. An important significance of this approach is the interaction of evanescent field with the propagating modes of radiation. Since the evanescent field is a decaying non-radiative field, it cannot be measured directly similar to the case with frustration of total internal reflection. Therefore, coupling techniques are required. An aperture with sub-wavelength scale dimension can generate evanescent field and couple the inhomogeneous field back to propagating modes. Enhancement of transmission of light through an aperture on a screen via evanescent field excitation has been demonstrated by many groups [10]. Notably, a sub-wavelength groove has been referred to as evanescent wave assist features. The evanescent wave enhancement in sub-wavelength grooves can be expressed as [11],

$$A(\lambda) = \left(1 + 2\sum_{j=1}^{N} \frac{\alpha d}{jP} \cos\left(\frac{2\pi}{\lambda} n_{eff}\, jP + \frac{\pi}{2}\right)\right)^2 \tag{4.1}$$

where λ is the wavelength of light, α is the evanescent wave amplitude relative to the incident light amplitude, n_{eff} is the effective index of refraction and P is the pitch of the sub-wavelength grooves.

The principle of evanescent wave interference lithography is that a shorter wavelength intensity pattern can be generated in the near-field of sub-wavelength grooves or prisms when two resonantly enhanced, evanescently decaying waves superimposed [12]. In the case of near-field techniques like EWIL, the recording medium should be kept in close proximity to either the high refractive index medium or the sub-wavelength mask to capture the exponentially decaying evanescent fields. That is, the amplitude of evanescent field decays exponentially with a distance away from the medium or mask/photoresist interface. While EWIL provides sub-wavelength scale resolution, it is still limited by low contrast and short exposure depth since the evanescent wave is a weakly decaying field. This limits the fabrication of high aspect ratio sub-wavelength structures.

4.3 Plasmonic Lithography

The issues associated with evanescent wave lithography can be overcome by using the concept of surface plasmon resonance [13]. In plasmonic lithography technique, high resolution interference patterns can be generated by the excitation of surface plasmons (SPs) on a thin metallic film that is generally put in contact with the photoresist layer, in order to expose the plasmon near-field [14]. The presence of enhanced field on the metal surface due to SPs can greatly enhance the light transmission and provides a novel method of near-field photolithography beyond the diffraction limit. Basically, surface plasmons are the enhanced evanescent field at the dielectric/metal interface, as a result the exposure depth and intensity contrast of the patterns generated via plasmonic lithography is much higher than that obtained with EWIL. In short, plasmonic lithography provides high resolution and high density, which can be used for the fabrication of high aspect ratio periodic nanostructures.

Surface plasmons are collective charge density oscillations occur at the interface between a conductor and a dielectric. Based on the nature of the mode, SPs can be divided in to two: a freely propagating electron density wave along a metal surface called surface plasmon polaritons (SPPs) and localized electron oscillations on a metal nanoparticle called localized surface plasmons (LSPs). Because of the unique optical properties of surface plasmons, a wide range of practical applications are possible such as biodetection at the single molecule level, high resolution optical imaging below the diffraction limit, guiding and manipulation of light at the nanoscale, negative index metamaterials, and enhanced optical transmission through subwavelength apertures [13].

The surface plasmon resonance-based nanoscale lithography, which is based on the enhancement of local field at the near-field, has theoretically and experimentally been demonstrated by many research groups [15–19]. The two important plasmonic lithography configurations proposed to pattern subwavelength nanostructures are metal mask-based configuration and Kretschmann configuration, which are respectively based on grating coupling and prism coupling technique of SPPs excitation. The plasmonic mask-based configuration at UV wavelength range has a great interest in plasmonic lithography to pattern subwavelength features using conventional near UV light sources, because of the principle of extraordinary optical transmission of the SPPs through the subwavelength aperture [15]. However, the plasmonic mask-based technique requires the fabrication of a fine-period mask grating or hole array, therefore they are found to be not cost-effective.

Kretschmann-Raether geometry is a well-known and efficient method for the excitation of surface plasmons and thus for plasmonic lithography [20]. This configuration has widely been used for many applications especially for SPR sensors, even in the commercial plasmonic biosensors. This is a two-layer configuration in which high refractive index prism (e.g. BK7) is in direct optical contact with a thin metal film (Au or Ag). Importantly, this is a maskless surface plasmon interference nanolithography configuration, which is based on the prism coupling principle of surface plasmon excitation. The prism coupling technique is considered to be one of the best SPR couplers because by using a high refractive index prism in Kretschmann–Raether geometry, it can achieve momentum matching condition in a very efficient way. In addition, large area patterning is possible using this geometry. More importantly, prism-based plasmonic lithography configuration has a potential to operate at different wavelengths to realize high resolution periodic features, however, the major technical challenge is to find a proper metal for SPP excitation at shorter wavelengths. Researchers are very active in this area to develop cost effective plasmonic lithography systems for the patterning of high throughput, good aspect ratio, and high-resolution nanoscale features.

4.4 Theoretical Analysis of Surface Plasmon Interference

As mentioned above, prism-based coupling is the efficient method to excite surface plasmon polaritons, performed with the evanescent field generated by attenuated total internal reflection principle. A schematic representation of SP interference lithography technique based on prism coupling method is illustrated in Fig. 4.1a. As can be seen, the upper layer is a high refractive index isosceles triangle prism, and the coated thin metal layer at the bottom surface of the prism, is in direct contact with a photoresist layer.

Surface plasmon dispersion relation is given by,

$$k_{SP} = k_x = \frac{\omega}{c}\sqrt{\frac{\varepsilon_m \varepsilon_d}{\varepsilon_m + \varepsilon_d}} \tag{4.2}$$

where ω is the excitation frequency, c is the speed of light in vacuum, ε_m is the complex dielectric constant of metal and ε_d is the dielectric constant of dielectric layer. The reflected light recorded in the far-field has a minimum value at the resonance angle when the surface plasmon resonance condition is satisfied. The resonance also

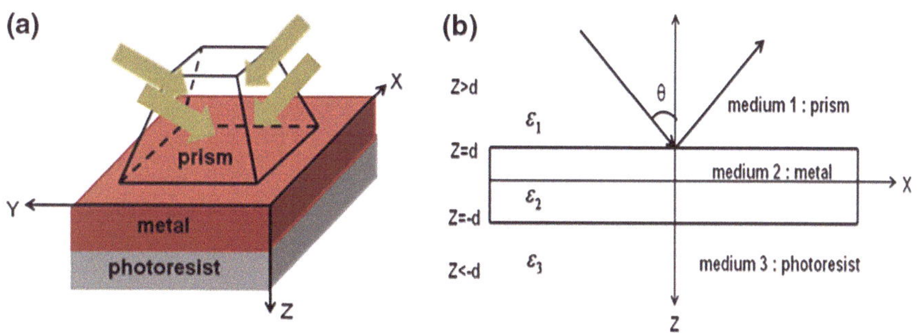

Fig. 4.1 **a** Schematic representation of multiple beams SP interference lithography configuration. **b** Schematic of considered three-layer system. Reproduced with permission from Ref. [21]

depends on the characteristics of the dielectric layer. Equation (4.2) expressed in terms of surface plasmon resonance angle (θ_{SP}) is given by,

$$k_x = k_{SP} = \frac{\omega}{c}\sqrt{\varepsilon_d}\sin\theta_{SP} = k_0 n_d \sin\theta_{SP} \tag{4.3}$$

The parameters such as incidence angle, wavelength of light and the dielectric constant of metal and dielectric layers are extremely important to satisfy the SPR resonance. At the surface plasmon resonance, the wavevector of SPPs becomes significantly larger than that of the incident light wavevector when the real parts of ε_m approaches-ε_d. On the other hand, the wavelength of excited SPPs is always shorter than the illumination light wavelength when this condition is satisfied.

We provide a detailed theoretical framework for multiple beams surface plasmon interference by considering a three-layer configuration [21]. The configuration shown in Fig. 4.1b is in the (x, z) plane and is identically represented in the (y, z) plane. We assume that the optical contact at prism/metal and metal/photoresist interfaces are ideal. The interface at the prism/metal or metal/photoresist is represented at $z = d$ and $z = -d$ respectively, and ε_1, $\varepsilon_2 = \varepsilon_2' + i\varepsilon_2''$ and ε_3 are the dielectric constant of prism, metal and photoresist layer, respectively.

One-dimensional periodic features (grating lines) can be generated on the photoresist (medium 3) when two p-polarized plane waves with equal incidence angle and amplitude are incident at the prism/metal interface. In fact, the surface plasmons can propagate in both positive and negative directions of the x-axis. For this case, the Maxwell's curl equations are given by,

$$\begin{aligned} \nabla \times H &= \frac{\partial D}{\partial t} \\ \nabla \times E &= -\frac{\partial B}{\partial t} \end{aligned} \tag{4.4}$$

In terms of harmonic time dependence $\left(\frac{\partial}{\partial t} = -i\omega\right)$, the set of coupled equations in Eq. (4.4) can be rewritten as

$$\begin{aligned} \frac{\partial E_z}{\partial y} - \frac{\partial E_y}{\partial z} &= i\omega\mu_0 H_x \\ \frac{\partial E_x}{\partial z} - \frac{\partial E_z}{\partial x} &= i\omega\mu_0 H_y \\ \frac{\partial E_y}{\partial x} - \frac{\partial E_x}{\partial y} &= i\omega\mu_0 H_z \\ \frac{\partial H_z}{\partial y} - \frac{\partial H_y}{\partial z} &= -i\omega\varepsilon\varepsilon_0 E_x \\ \frac{\partial H_x}{\partial z} - \frac{\partial H_z}{\partial x} &= -i\omega\varepsilon\varepsilon_0 E_y \end{aligned}$$

$$\frac{\partial H_y}{\partial x} - \frac{\partial H_x}{\partial y} = -i\omega\varepsilon\varepsilon_0 E_z \tag{4.5}$$

In the (*x*, *z*) plane of the medium 1(*z* > d), the electromagnetic fields of a single beam can be obtained as follows:

The coupled equations in (4.5) simplifies to get the following equations when the electromagnetic field propagation is along the *x*-direction and homogeneity in the *y*-direction, that is $\left(\frac{\partial}{\partial x} = ik_x\right)$ and $\left(\frac{\partial}{\partial y} = 0\right)$

$$\begin{aligned}
&\frac{\partial E_y}{\partial z} = -i\omega\mu_0 H_x \\
&\frac{\partial E_x}{\partial z} - ik_x E_z = i\omega\mu_0 H_y \\
&ik_x E_y = i\omega\mu_0 H_z \\
&\frac{\partial H_y}{\partial z} = i\omega\varepsilon\varepsilon_0 E_x \\
&\frac{\partial H_x}{\partial z} - ik_x H_z = -i\omega\varepsilon\varepsilon_0 E_y \\
&ik_x H_y = i\omega\varepsilon\varepsilon_0 E_z
\end{aligned} \tag{4.6}$$

Note that the electric component has no *y* dependence when *p*-polarized plane waves propagate in the *x*-direction. As a result, the field components are (E_x, H_y, E_z).

For the medium 1, the electric field components can be obtained from Eq. (4.6) as,

$$E_{x1}(x, z) = -i\frac{1}{\omega\varepsilon_1\varepsilon_0}\frac{\partial H_{y1}}{\partial z} \tag{4.7}$$

$$E_{z1}(x, z) = -\frac{k_x}{\omega\varepsilon_1\varepsilon_0}H_{y1} \tag{4.8}$$

The magnetic field component is given by,

$$H_{y1}(x, z) = A_1 e^{ik_x x} e^{k_{z1} z} \tag{4.9}$$

By substituting Eq. (4.9) in Eqs. (4.7) and (4.8), we get,

$$E_{x1}(x, z) = -iA_1\frac{k_{z1}}{\omega\varepsilon_1\varepsilon_0}e^{ik_x x}e^{-k_{z1} z} \tag{4.10}$$

$$E_{z1}(x, z) = -A_1\frac{k_x}{\omega\varepsilon_1\varepsilon_0}e^{ik_x x}e^{k_{z1} z} \tag{4.11}$$

where $k_{z1} = \frac{2\pi}{\lambda}\left(\frac{\varepsilon_2'^2}{\varepsilon_2'+\varepsilon_1}\right)^{1/2}$ and $k_x = \frac{2\pi}{\lambda}\left(\frac{\varepsilon_2'\varepsilon_1}{\varepsilon_2'+\varepsilon_1}\right)^{1/2}$

In Eqs. (4.10) and (4.11), ω and ε_0 are the excitation frequency and free space dielectric constant, respectively. In the medium 3 ($z < -$d), the corresponding field components are given by,

$$\begin{aligned} H_{y3}(x,z) &= A_3 e^{ik_x x} e^{-k_{z3} z} \\ E_{x3}(x,z) &= iA_3 \frac{k_{z3}}{\omega\varepsilon_3\varepsilon_0} e^{ik_x x} e^{-k_{z3} z} \\ E_{z3}(x,z) &= -A_3 \frac{k_x}{\omega\varepsilon_3\varepsilon_0} e^{ik_x x} e^{-k_{z3} z} \end{aligned} \tag{4.12}$$

where $k_{z3} = \frac{2\pi}{\lambda}\left(\frac{\varepsilon_2^{'2}}{\varepsilon_2^{'}+\varepsilon_3}\right)^{1/2}$ and $k_x = \frac{2\pi}{\lambda}\left(\frac{\varepsilon_2^{'}\varepsilon_3}{\varepsilon_2^{'}+\varepsilon_3}\right)^{1/2}$

For the metal medium 2 ($-$d < z < d), the coupled electromagnetic fields are given by,

$$\begin{aligned} H_{y2}(x,z) &= A_2 e^{ik_x x} e^{k_{z2} z} + A_2' e^{ik_x x} e^{-k_{z2} z} \\ E_{x2}(x,z) &= -iA_2 \frac{k_{z2}}{\omega\varepsilon_2\varepsilon_0} e^{ik_x x} e^{k_{z2} z} + iA_2' \frac{k_{z2}}{\omega\varepsilon_2\varepsilon_0} e^{ik_x x} e^{-k_{z2} z} \\ E_{z2}(x,z) &= A_2 \frac{k_x}{\omega\varepsilon_2\varepsilon_0} e^{ik_x x} e^{k_{z2} z} + A_2' \frac{k_x}{\omega\varepsilon_2\varepsilon_0} e^{ik_x x} e^{-k_{z2} z} \end{aligned} \tag{4.13}$$

where $k_{z2} = \frac{2\pi}{\lambda}\left(\frac{\varepsilon_1^2}{\varepsilon_2^{'}+\varepsilon_1}\right)^{1/2}$ or $k_{z2} = \frac{2\pi}{\lambda}\left(\frac{\varepsilon_3^2}{\varepsilon_2^{'}+\varepsilon_3}\right)^{1/2}$

In fact, the electromagnetic field underneath the metal layer (medium 3) is interested in surface plasmon interference lithography. Thus, the electric field components in terms of surface plasmons propagating along the x-axis with the components, $H_{SP} = (0, H_{y,SP}, 0)$ and $E_{SP} = (E_{x,SP}, 0, E_{z,SP})$ are given by,

$$H_{y,SP}(x,z) = A_3 e^{ik_x x} e^{-k_{z3} z} \tag{4.14}$$

$$E_{x,SP}(x,z) = iA_3 \frac{k_{z3}}{\omega\varepsilon_3\varepsilon_0} e^{ik_x x} e^{-k_{z3} z} \tag{4.15}$$

$$E_{z,SP}(x,z) = -A_3 \frac{k_x}{\omega\varepsilon_3\varepsilon_0} e^{ik_x x} e^{-k_{z3} z} \tag{4.16}$$

The electric field equations can be further simplified to,

$$E_{x,SP}(x,z) = iA_3 \frac{k_{z3}}{\omega\varepsilon_3\varepsilon_0} e^{-k_{z3} z} \cos k_x x \tag{4.17}$$

$$E_{z,SP}(x,z) = A_3 \frac{k_x}{i\omega\varepsilon_3\varepsilon_0} e^{-k_{z3} z} \sin k_x x \tag{4.18}$$

From Eqs. (4.17) and (4.18), it is clear that the amplitude of E_x and E_z components decrease exponentially with $\exp(-k_{z3}z)$. This evidences that the SP electric field components are exponentially decaying fields and the amplitude ratio of SP field components is given by,

$$\frac{|E_{z,SP}|}{|E_{x,SP}|} = \left|\frac{k_x}{k_z}\right| \approx \left|\frac{\mathrm{Re}(\varepsilon_2)}{\varepsilon}\right| \tag{4.19}$$

It is evident from Eq. (4.19) that the field component $E_{z,SP}(x, z)$ dominates over the component, $E_{x,SP}(x, z)$. As a result, the SP interference term is determined by $E_{z,SP}(x, y, z)$. Therefore, the intensity distributions of the interference between two counter propagating surface plasmon waves are given by,

$$I_{x,SP} \propto \left\langle \left(E_{x,SP} + E'_{x,SP}\right) \cdot \left(E_{x,SP} + E'_{x,SP}\right) \right\rangle \tag{4.20}$$

$$I_{z,SP} \propto \left\langle \left(E_{z,SP} + E'_{z,SP}\right) \cdot \left(E_{z,SP} + E'_{z,SP}\right) \right\rangle \tag{4.21}$$

where $E'_{x,SP}(x, z)$ and $E'_{z,SP}(x, z)$ are the $E_{x,SP}$ and $E'_{z,SP}(x, z)$ field components in the negative (x, z) direction.

The propagation of surface plasmons in both positive and negative directions of the x- and y-axis is possible when four counter propagating p-polarized plane waves with identical incidence angle and amplitude are incident at the prism/metal interface. In the case of four beams interference, two-dimensional periodic dot arrays with square lattice symmetry can be generated on the photoresist. In this case, the electromagnetic field components in the (y, z) plane, in addition to (x, z) plane needs to be find out. Therefore, the electromagnetic fields of a single beam in the (y, z) plane of the medium $3(z < -\mathrm{d})$ can be obtained as follows,

Here, the electromagnetic field propagation is along the y-direction and homogeneity in the x-direction, therefore $\left(\frac{\partial}{\partial y} = ik_y\right)$ and $\left(\frac{\partial}{\partial x} = 0\right)$. By applying the same procedure described above, the electric field component in the (y, z) plane of the medium 3 can be obtained as,

$$E_{y3}(y, z) = i\frac{1}{\omega\varepsilon_3\varepsilon_0}\frac{\partial H_{x3}}{\partial z} \tag{4.22}$$

$$E_{z3}(y, z) = \frac{k_y}{\omega\varepsilon_3\varepsilon_0}H_{x3} \tag{4.23}$$

The corresponding magnetic field component is given by,

$$H_{x3}(y, z) = A_3 e^{ik_y y} e^{-k_{z3}z} \tag{4.24}$$

The SP field components in the y-direction of SP propagation are given by, $H_{SP} = (H_{x,SP}, 0, 0)$ and $E_{SP} = (0, E_{y,SP}, E_{z,SP})$. By substituting Eq. (4.24) in Eqs. (4.23) and (4.22), we get,

$$E_{y,SP}(y,z) = -iA_3 \frac{k_{z3}}{\omega\varepsilon_3\varepsilon_0} e^{ik_y y} e^{-k_{z3}z} \tag{4.25}$$

$$E_{z,SP}(y,z) = A_3 \frac{k_x}{\omega\varepsilon_3\varepsilon_0} e^{ik_y y} e^{-k_{z3}z} \tag{4.26}$$

On the other hand, the SP electric field components in the (y, z) plane is given by,

$$E_{y,SP}(y,z) = -iA_3 \frac{k_{z3}}{\omega\varepsilon_3\varepsilon_0} e^{-k_{z3}z} \cos k_y y \tag{4.27}$$

$$E_{z,SP}(y,z) = -A_3 \frac{k_y}{i\omega\varepsilon_3\varepsilon_0} e^{-k_{z3}z} \sin k_y y \tag{4.28}$$

For four beams SP interference, the total electric field components can be obtained by adding Eqs. (4.17), (4.18), (4.27) and (4.28), and which are given by,

$$E_{x,SP}(x,z) = iA_3 \frac{k_{z3}}{\omega\varepsilon_3\varepsilon_0} e^{-k_{z3}z} \cos k_x x \tag{4.29}$$

$$E_{y,SP}(y,z) = -iA_3 \frac{k_{z3}}{\omega\varepsilon_3\varepsilon_0} e^{-k_{z3}z} \cos k_y y \tag{4.30}$$

$$E_{z,SP}(x,y,z) = A_3 \frac{k_{SP}}{i\omega\varepsilon_3\varepsilon_0} e^{-k_{z3}z} (\sin k_x x - \sin k_y y) \tag{4.31}$$

The amplitude term in Eq. (4.30) can take $k_x \approx k_y = k_{SP}$

As can be seen, the amplitude of E_x, E_y and E_z components decreases exponentially with $\exp(-k_{z3}z)$, which is an indication that the SP electric field components are exponentially decaying fields. The amplitude ratio of SP field component is given by,

$$\frac{|E_{z,SP}|}{|E_{x,SP}|} = \frac{|E_{z,SP}|}{|E_{y,SP}|} = \left|\frac{k_{SP}}{k_{z3}}\right| \approx \left|\frac{\mathrm{Re}(\varepsilon_2)}{\varepsilon_3}\right| \tag{4.32}$$

It shows that the field component $E_{z,SP}(x, y, z)$ dominates over other two components, $E_{x,SP}(x, , z)$ and $E_{y,SP}(y, z)$. As a result, the SP interference term is determined by $E_{z,SP}(x, y, z)$. In the case of four beams interference, the field components $E_{x,SP}(x, z)$ and $E_{y,SP}(y, z)$ generate equal intensity one-dimensional periodic features on the (x, z) and (y, z) plane respectively, whereas $E_{z,SP}(x, y, z)$ field component generates high intensity two-dimensional periodic features on the (x, y, z) plane. The intensity distribution of the interference between four counter propagating surface plasmon waves is given by,

$$I_{z,SP} \propto \left\langle \left(E_{z,SP}(x,y,z) + E'_{z,SP}(x,y,z)\right) \cdot \left(E_{z,SP}(x,y,z) + E'_{z,SP}(x,y,z)\right)\right\rangle \tag{4.33}$$

where $E'_{z,SP}(x, y, z)$ is the $E_{z,SP}$ field component in the negative (x, y, z) direction.

On another aspect, if ψ_1 and ψ_2 are the two SP waves, then their superposed interference pattern can be expressed mathematically as [22],

$$|\psi_1 + \psi_2|^2 = 2|\psi_0|^2\left[1 + \cos\left(\frac{4\pi\sqrt{\varepsilon_d}\sin\theta_{Sp}}{\lambda}\right)x\right] \tag{4.34}$$

where ψ_0 is the incident wave, ε_d is the dielectric function of photoresist layer, λ is the illumination wavelength and θ_{SP} is the SP resonant angle.

The SP interference fringe separation can be expressed as [22],

$$\Delta x = \frac{\lambda}{2\sqrt{\varepsilon_d}\sin\theta_{SP}} \tag{4.35}$$

4.5 Numerical Analysis of Surface Plasmon Interference

We further performed numerical simulations of multiple beams surface plasmon interference, using finite difference time domain (FDTD) method [23, 24]. We considered the lithography configuration shown in Fig. 4.1a for the numerical simulations. In the simulation model, the interface between the metal and photoresist is represented at $z = 0$, as shown in Fig. 4.2. The boundary conditions used in the simulations are: perfectly matched layer (PML) along the z-axis, which is perpendicular to the propagation direction of the light and periodic boundary conditions along other two axis. We used a fine mesh size of 1 nm along the three axis. A *p*-polarized plane wave with a wavelength of 364 nm is incident on the prism surface. The refractive index and thickness of photoresist layer is set to be 1.7 and 300 nm, respectively. We investigated the behavior of surface plasmon interference patterns by using Al and Ag metal films. The thickness of metal film is set to be 50 nm. At the excitation wavelength of 364 nm, used complex dielectric constant of Al and Ag is $-19.42 + 3.606i$ and $-2.5575 + 0.5989i$, respectively.

Figure 4.2 shows the intensity field distributions of the interference pattern generated on the photoresist layer when two counter propagating *p*-polarized plane waves incident on the prism/metal interface at the SPR resonant angle. In Fig. 4.2, z-axis is the decay direction and metal/photoresist interface is the origin of the decay direction. The intensity distribution ($I = |E|^2$) obtained with Al is shown in Fig. 4.2a for E_x component and in Fig. 4.2b for E_z component. The corresponding E_x and E_z components generated with Ag metal film is shown in Fig. 4.2c, d, respectively. It is clear that both field components generate one-dimensional periodic patterns (grating lines) on the photoresist. As expected, the normalized intensity of both field components is highest at the metal/photoresist interface and decay exponentially along the z-direction. The magnitude of E_z field component at the origin is 45 for Al and 21 for Ag. Note that the intensity values are normalized with respect to the incident light intensity.

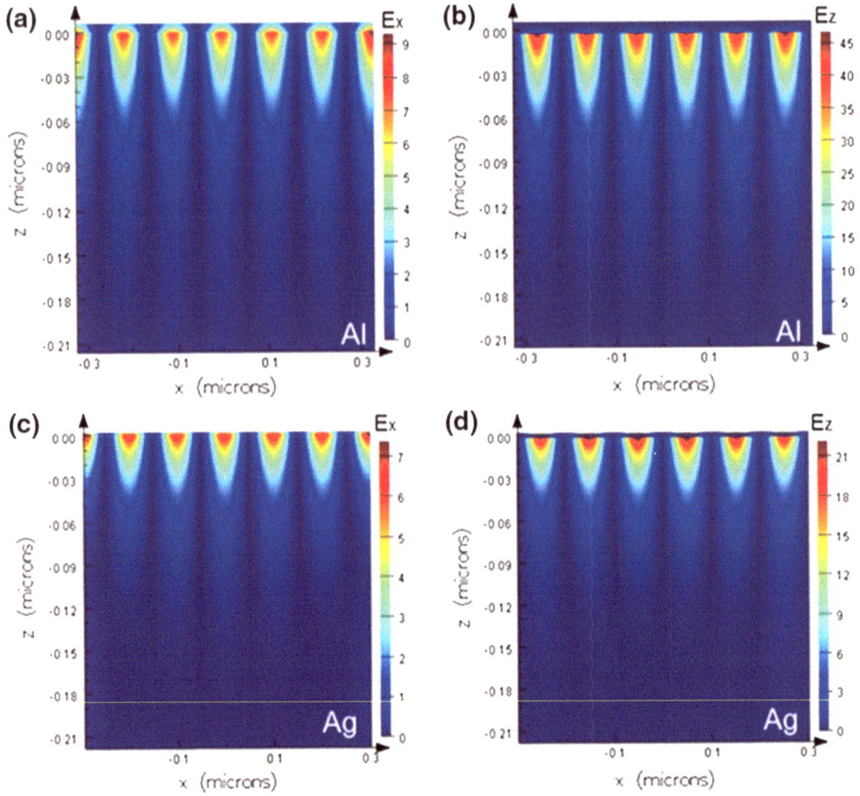

Fig. 4.2 Electric intensity field distribution of the interference pattern generated on the photoresist layer. **a** Al-E_x, **b** Al-E_z, **c** Ag-E_x, and **d** Ag-E_z. Reproduced with permission from Ref. [28]

In the case of multiple beams SP interference simulations, we considered four equal amplitude *p*-polarized plane waves incident on the prism. As shown in Fig. 4.3, two-dimensional sub-wavelength periodic patterns are generated. The E_z intensity component of two-dimensional periodic patterns generated on the (x, y) plane using Al and Ag is shown in Fig. 4.3a, b, respectively. It shows a periodic dot array pattern with square lattice symmetry. In this case also the magnitude of E_z field component is higher for Al. Therefore, it can be concluded that Al metal provides higher intensity patterns on the recording medium as compared to Ag, which could result higher contrast periodic patterns in experiments. It is due to the fact that the resonant surface plasmon wavelength of Al locates in the UV wavelength range. A key finding in the simulation is that the E_z field component dominates over E_x field component in both two and four beams SP interference configuration. It correlates very well with our theoretical predictions.

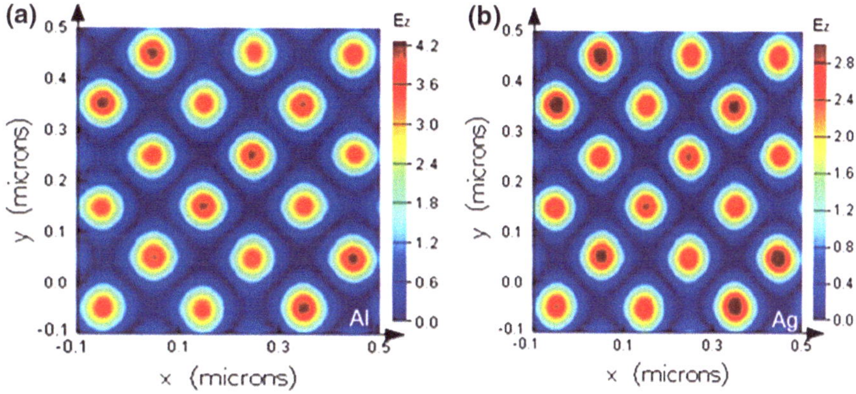

Fig. 4.3 E_Z-intensity field distribution along the (x, y) plane using **a** Al and **b** Ag

4.6 Nanopatterning Based on Multiple Beams Surface Plasmon Interference

A home-built interferometric lithography set-up has been used for the experimental realization of multiple beams SP interference. The main components of the set-up are: a continuous-wave argon-ion laser with wavelength 364 nm, an electronic shutter, a cross-phase grating and polarizers [25–28]. In Fig. 4.4, we illustrate the schematic diagram of the experimental setup. As can be seen, an electronic shutter (Newport, model 845 HP) controlled continuous wave argon-ion laser (Spectra Physics, 2065-7S) with an illumination wavelength of 364 nm is normally incident on a custom made cross-phase diffraction grating (Ibsen Photonics Pte. Ltd.). The cross-phase grating was designed in such a way that it could suppress the high intensity 0th order beam and generates high diffraction efficiency (17%) four first order beams. A linear polarizer placed on the custom fabricated rotating mount, was positioned in the optical path of each diffracted beam, for controlling the polarization of the light beam. During the exposure, the 0th order beam was blocked with a hard stop and four counter propagating first order beams were incident on the sample. The first order diffraction beams were aligned and directed in such a way that they are made to incident at 45° on the sample. The average power density of incident beams measured on the sample is 334.35 mW/cm^2. For two beams SP interference, two counter propagating first order beams were used, however, four first order beams were used when four beams interference was employed.

In the prism configuration, we used a flat top high refractive index 60° triangular prism. The flat top surface is purposefully designed to reflect any stray light incident on the prism. Initially, a thin metal film (Al and Ag) of thickness 50 nm was deposited on the bottom surface of the prism via physical vapor deposition technique. In order to record the SP interference patterns, a thin positive tone photoresist (AZ7220) was spin coated on the silicon substrate with a sample dimension of 20 mm × 20 mm.

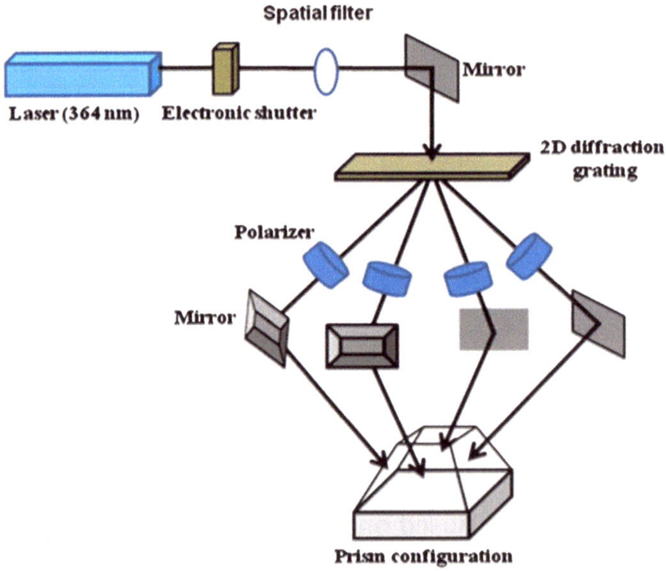

Fig. 4.4 Schematic representation of multiple beams SP interference lithography experimental setup. Reproduced from Ref. [26], with the permission of the American Vacuum Society

AZ7200 photoresist was diluted with PGMEA (EBR solvent, AZ Electronic Material) at different proportions and spin coated at 1500 rpm to obtain thin photoresist layer. As prepared samples were kept in intimate optical contact with the metal coated prism using index matching liquid (Cargille oil). The exposed samples were developed in AZ300MIF.

Firstly, we discuss the results of two beams SP interference lithography. In this case, two counter propagating first-order beams coming out of the cross-phase grating were allowed to incident on the prism. The measured thickness of photoresist layer was ~100 nm. Initially, Al film was deposited on the bottom surface of a NLAF 36 glass prism with a refractive index of 1.865. The calculated SP resonance angle at the prism/ Al interface is 52.3°. The sample was exposed with a time duration of 15 s and then it was immediately developed in AZ300MIF solution with a time duration of 2 min. We then used a tapping-mode atomic force microscope (AFM) with high lateral resolution (5 nm) to analysis the topography of the recorded features on the photoresist layer. In Fig. 4.5a, we show the AFM image of the SP interference pattern recorded on the photoresist using two beams interference. For the optimized exposure parameters, large-area uniform sub-wavelength periodic grating lines were obtained.

We repeated the experiments by coating Ag film on bottom surface of the prism. Here, we used NLAK 8 glass prism with a refractive index of 1.745, therefore the calculated SP resonance angle at the prism/Ag interface is 54°. In this case, a lower exposure time of 1 s and a development time of 2 min were used. The AFM image

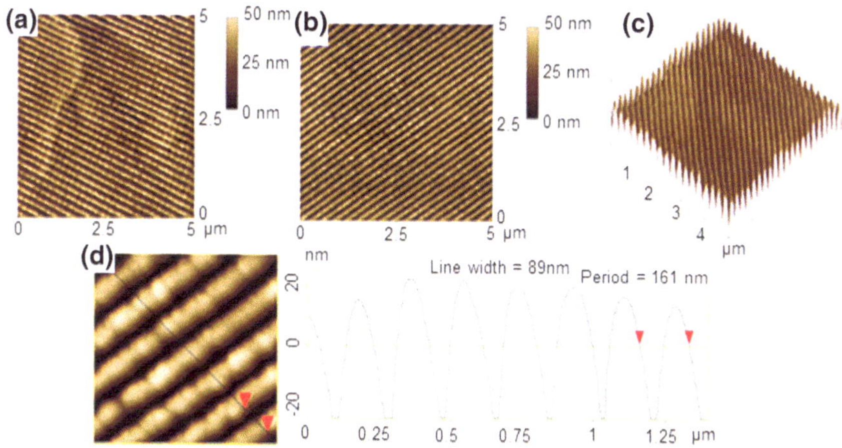

Fig. 4.5 Fabricated large-area grating lines on the photoresist film **a** 2D AFM image using Al at an exposure time of 15 s, **b** 2D AFM image using Ag at an exposure time of 1 s. **c** Corresponding 3D AFM image using Ag. **d** Sectional analysis along the line drawn; periodic patterns with a minimum line width of 89 nm and a period of 161 nm are obtained (for Ag as the metal medium). Reproduced with permission from Ref. [25]

of exposed pattern transferred on the photoresist layer using Ag film is shown in Fig. 4.5b for 2D image and the corresponding 3D image is shown in Fig. 4.5c. A sectional analysis of the AFM image is shown in Fig. 4.5d. Based on the topography analysis, we concluded that the grating lines with an average fringe separation of 161 nm and an average line width of 89 nm were patterned on the photoresist layer using both Al and Ag thin films.

Now, we discuss the patterning of 2D sub-wavelength periodic features on the recording medium using multiple beams SP interference lithography. To generate interference of multiple counter propagating SP waves as a lithographic technique to print periodic 2D features, four first-order beams of the 2D diffraction grating were used. The obtained results using thin films of Al and Ag are shown in Fig. 4.6. A 2D and 3D AFM topography image of the exposed dot array pattern transferred onto the photoresist layer using Al film is shown in Fig. 4.6a, c, respectively. The corresponding AFM sectional analysis of the exposed dot array on the photoresist film is shown in Fig. 4.6d. It is confirmed that large-area periodic dot array pattern with a minimum spot size of 93 nm and an average period of 175 nm was recorded on the photoresist. A 2D AFM image of the recorded dot array pattern on the photoresist layer using Ag film is shown in Fig. 4.6b.

The observed non-uniformity of the exposed dot array patterns could be due to the following reasons: (i) laser intensity fluctuation, (ii) its dependence on the exposure duration, power density and development time, and (iii) the possible sample drift while doing AFM measurement. We found that the obtained resolution and period of the pattern in two- and four-beams SP interference lithography using both metal films are almost the same. However, there must be some difference in exposure depth

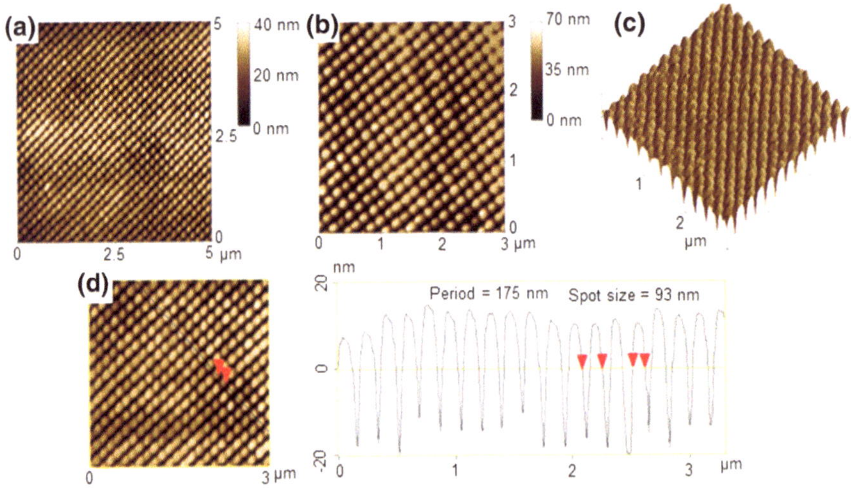

Fig. 4.6 Fabricated large-area periodic dot array on the photoresist film at an exposure time of 1 s. **a** 2D AFM image using Al at an exposure time of 15 s, **b** 2D AFM image using Ag at an exposure time of 1 s and **c** corresponding 3D AFM image using Al. **d** Sectional analysis along the line drawn; periodic patterns with a minimum spot size of 93 nm on a 175 nm period were obtained (for Al as the metal medium). Reproduced from [26], with the permission of the American Vacuum Society

and intensity contrast as evident from the numerical simulations. We investigate this feature by spin coating thick photoresist layer. The estimated period (using Eq. 4.35) of two beams SP interference lithography for Ag and Al metal film is 164 and 160 nm, respectively. The calculated periods are comparable with the experimentally obtained period of Ag and Al.

To analyze the metal film dependence on the SP interference, we then repeated two and four beams SP interference lithography using thick photoresist film. For this purpose, we spin coated a comparatively thick photoresist layer of thickness ~240 nm on the silicon wafer. By using Al and Ag thin films, the obtained results of two- and four-beams SP interference lithography are shown in Fig. 4.7. Here, the samples were exposed for a duration of 25 s for Al and 3 s for Ag and the development time was set to be 2 min. The generated period and resolution of the periodic structures are almost same for Al and Ag metal films, however, the height of the feature is slightly higher for Al thin film (200 nm for two beams SPIL and 180 nm for four beams SPIL). It indicates that SPIL with Al metal gives good exposure depth and better intensity contrast when we use an UV illumination wavelength of 364 nm, which correlate well with our numerical simulations.

For 364 nm wavelength laser excitation, a comparison between three interference lithography techniques such as conventional laser interference, evanescent wave interference and surface plasmon interference are tabulated in the Table 4.1. In contrast to conventional laser interference lithography, near-field interference lithography techniques such as EWIL and SPIL provide much higher resolution.

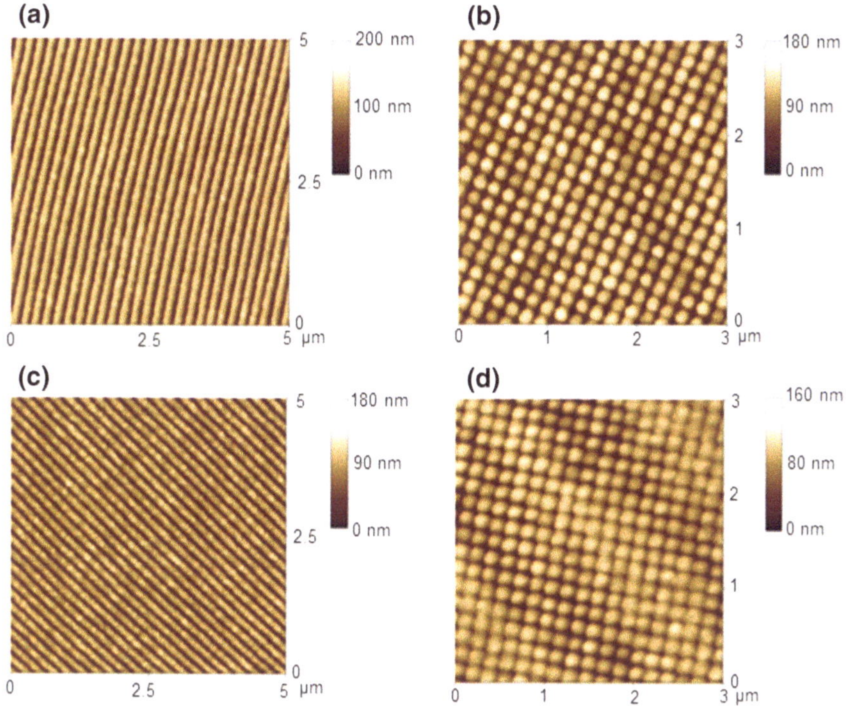

Fig. 4.7 Two and four beams SP interference lithography using thick photoresist film. **a** 1D grating using Al film, **b** 2D grating using Al film, **c** 1D grating using Ag film and **d** 2D grating using Ag film. Reproduced with permission from Ref. [25]

Moreover, SPIL provides long exposure depth and good intensity contrast as compared to EWIL, which can be used for the fabrication of high aspect ratio periodic features. In addition, SPIL can provide high aspect ratio features close to LIL by choosing suitable metal at a specific wavelength.

It is important to note that the prism-based configuration is the cost-effective plasmonic lithography technique for fabricating large-area nanostructures. In particular, surface plasmon interference lithography is an efficient, cost-effective, maskless patterning technique for the fabrication of high resolution, good intensity contrast and high aspect ratio periodic features. The symmetry of the periodic nanostructures can be controlled by choosing number of courter propagating beams. For example, periodic nanostructures with hexagonal lattice symmetry can be achieved by using single exposure three beams SP interference. Future investigations would be along this direction. It is also possible to further improve the aspect ratio and scalability of the periodic sub-wavelength pattern by using the concept of long-range surface plasmon (LRSP) interference [29].

Table 4.1 Comparison chart of various interference lithography techniques at 364 nm wavelength

Interference lithography	Two beams interference			Four beams interference			Theoretically estimated period (nm)
	Average period (nm)	Average resolution (nm)	Maximum height (nm)	Average period (nm)	Average resolution (nm)	Maximum height (nm)	
LIL (using p-polarization) [27]	258.8	136.7	300	241.5	131	300	257
EWIL (using p-polarization) [27]	185.5	88	100	170	88	100	182
SPIL (using Ag)	164	84.5	180	166	88	160	160
SPIL (using Al)	164	84.5	200	166	88	180	164

References

1. Okazaki S (1991) Resolution limits of optical lithography. J Vac Sci Technol B 9:2829
2. Ji R, Lee W, Scholz R, Gosele U, Nielsch K (2006) Templated fabrication of nanowire and nanoring arrays based on interference lithography and electrochemical deposition. Adv Mater 18:2593
3. Fan S, Zhang S, Malloy KJ, Brueck SRJ (2005) Large-area, infrared nanophotonic materials fabricated using interferometric lithography. J Vac Sci Technol, B 23:2700
4. Moon JH, Ford J, Yang S (2006) Fabricating three-dimensional polymeric photonic structures by multi-beam interference lithography. Polym Adv Technol 17:83
5. Lee W, Ji R, Ross C, Gosele U, Nielsch K (2006) Wafer-scale Ni imprint stamps for porous alumina membranes based on interference lithography. Small 2:978
6. de Boor J, Geyer N, Gosele U, Schmidt V (2009) Three-beam interference lithography: upgrading a Lloyd's interferometer for single-exposure hexagonal patterning. Opt Lett 34:1783
7. Milster T, Jo JS, Hirota K (1999) Roles of propagating and evanescent waves in solid immersion system. Appl Opt 38:5046
8. Blaikie RJ, McNab SJ (2001) Evanescent interferometric lithography. Appl Opt 40:1692
9. Lezec HJ, Thio T (2004) Diffracted evanescent wave model for enhanced and suppressed optical transmission through subwavelength hole arrays. Opt Exp 12:3629–3651
10. Lafferty NV, Zhou J, Smith BW (2007) Mask enhancement using an evanescent wave effect. Proc SPIE 6520:652041
11. Ohdaira Y et al (2006) Fabrication of surface relief subwavelength gratings on azo dye thin films utilizing an interference of evanescent waves. J Opt A Pure Appl Opt 8:213
12. Chua JK, Murukeshan VM (2009) UV laser-assisted multiple evanescent waves lithography for near-field nanopatterning. Micro Nano Lett 4:210
13. Brongersma ML, Kik PG (2007) Surface plasmon nanophotonics. Springer, Berlin
14. Gramotnev DK, Bozhevolnyi SI (2010) Plasmonics beyond the diffraction limit. Nat Photon 4:83
15. Srituravanich W, Fang N, Sun C, Luo Q, Zhang X (2004) Plasmonic nanolithography. Nano Lett 4:1085
16. Luo X, Ishihara T (2004) Surface plasmon resonant interference nanolithography technique. Appl Phys Lett 84:4780
17. Shao DB, Chen SC (2005) Numerical simulation of surface plasmon assisted nanolithography. Opt Exp 13:6964
18. Srituravanich W, Durant S, Lee H, Sun C, Zhang X (2005) Deep subwavelength nanolithography using localized surface plasmon modes on planar silver mask. J Vac Sci Technol B 26:2636
19. Seo S, Kim HC, Ko H, Cheng M (2007) Subwavelength proximity nanolithography using a Plasmonic lens. J Vac Sci Technol B 25:227
20. Guo X, Du J, Guo Y, Yao J (2006) Large-area surface-plasmon polariton interference lithography. Opt Lett 31:2613
21. Sreekanth KV, Murukeshan VM (2011) Multiple beams surface plasmon interference generation: a theoretical analysis. Opt Commun 284:2042–2045
22. Passian A, Wig A, Lereu AL, Evans PG, Meriaudeau F, Thundat T, Ferrell TL (2004) Probing large area surface plasmon interference in thin metal films using photon scanning tunneling microscopy. Ultramicroscopy 100:429
23. Sreekanth KV, Murukeshan VM, Chua JK (2008) A planar layer configuration for surface plasmon interference nanoscale lithography. Appl Phys Lett 93:093103
24. Murukeshan VM, Sreekanth KV (2009) Excitation of gap modes in a metal particle-surface system for sub-30 nm plasmonic lithography. Opt Lett 34:845–847
25. Sreekanth KV, Murukeshan VM (2010) Large-area maskless surface plasmon interference for one-and two-dimensional periodic nanoscale feature patterning. J Opt Soc Am A 27:95–99
26. Sreekanth KV, Murukeshan VM (2010) Four beams surface plasmon interference nanoscale lithography for patterning of two-dimensional periodic feature. J Vac Sci Technol B 28:128–130

27. Sreekanth KV, Chua JK, Murukeshan VM (2010) Interferometric lithography for nanoscale feature patterning: a comparative analysis between laser interference, evanescent wave interference, and surface plasmon interference. Appl Opt 49:6710–6717
28. Sreekanth KV, Murukeshan VM (2010) Single exposure maskless plasmonic lithography for patterning of periodic nanoscale grating features. J Micro/Nanolith MEMS MOEMS 9:023007
29. Sreekanth KV, Ting Y (2013) Long range surface plasmons in a symmetric graphene system with anisotropic dielectrics. J Opt 15:055002

Part II
Development and Applications of Multilayered Hyperbolic Metamaterials

Hyperbolic metamaterials (HMMs) are an emerging research field over the past few years, and various material combinations have been used to engineer their dispersion relation in different spectral bands. In contrast to metamaterial designs rely on nanopatterning, especially in the optical frequencies, HMMs can easily be fabricated using conventional thin-film physical vapor deposition techniques, by depositing thin metal–dielectric multilayers in the effective medium regime. The extreme anisotropic permittivity of HMM represents a unique opportunity to realize effective bulk metastructures with extraordinary optical properties over a broad frequency range from ultraviolet to terahertz. In this part, we present the experimental realization of visible and near-infrared HMMs and their applications in biosensing, spontaneous emission enhancement, perfect light absorption, and super lensing. We further propose graphene and topological insulator-based tunable HMMs for realizing negative refraction in terahertz frequencies.

Chapter 5
Dielectric Singularities in Hyperbolic Metamaterials

In this chapter, a special propagation regime is described, occurring in particular HMMs with extreme parameters. These HMMs are designed to manifest a vanishing effective permittivity in the direction parallel to their surface plane while an ideally infinite one in the perpendicular direction. Such a unique feature allows the HMM to behave as a perfect near-field lens. As will be discussed in the chapter, all the high-frequency Fourier components of an object placed at the top of such a system are entirely preserved while traversing the HMM with a spatial resolution comparable to the period of the HMM. This innovative framework lays the foundations for more sophisticated scenarios discussed in the next chapters, involving suitably designed fluorescent materials, towards supercollimated nano-LASER emission and self-amplified perfect lensing.

5.1 Introduction

Hyperbolic Metamaterials (HMMs) are artificial nanostructures characterized by indefinite (hyperbolic) dispersion relation. They are usually composed of metallic and dielectric elements; whose interplay endows the overall metamaterial with remarkable anisotropic properties. Two geometries showing hyperbolic dispersion are of particular interest. The first one involves metallic nanowires immersed in a dielectric matrix, while the second consists of stacked metal/dielectric multilayers [1–4]. Due to its straightforward fabrication, the latter is the most studied. Multilayered HMMs have been used as the technological core in many applications from ultra-sensitive sensors to super-resolution imaging [5–12]. The Photonic Density of States (PDOS) in the proximity of an HMM can be finely engineered, opening to unexpected interaction possibilities with, for instance, weakly coupled fluorophores, whose lifetime and photoluminescence can be remarkably enhanced [13–16]. Thermal flux engineering has been achieved in these media, which have also been suggested as the near-field analog of a blackbody, and the occurrence of super-Planckian thermal energy transfer has been predicted [17, 18]. Noticeably, HMMs revealed themselves a stimulating

Sreekanth K. V. et al., *New Directions in Thin Film Nanophotonics*, Progress in Optical Science and Photonics 6, https://doi.org/10.1007/978-981-13-8891-0_5

platform for the investigation of sophisticated phenomena belonging to fields that are traditionally very far from them, configuring as a theoretical and experimental toolbox for exploring new frontiers in cosmology and astrophysics [19, 20].

5.2 Effective Medium Theory and HMMs Dispersion Relation

Since both the fundamental components of the HMMs are deeply subwavelength, the overall optical response of an HMM can be qualitatively predicted in the framework of the so-called *effective medium theory* (EMT) [21]. It is worth noticing that such an approximation completely neglects some crucial features typical of metal/dielectric multilayers, like the occurrence of ENZ (epsilon-near-zero) resonances in the so-called "*effective dielectric*" regime, which have recently attracted interest in the field of plasmonics [22, 23]. Moreover, the simple approximation provided by the EMT does not allow to consider the effect of the strong non-local response of these structures, so that many corrections have been introduced in this direction [24, 25]. Nonetheless, the EMT is still able to capture some specific remarkable feature of the wave propagation inside these metastructures, among which the so-called *Epsilon-Near-Zero-and-Pole* condition finds place, which will be the main topic of this chapter. Therefore, it is essential to lay the fundaments of this theory.

When stacking multiple metal/dielectric bilayers to create an HMM, a one-dimensional crystal is formed. Due to this particular geometric arrangement, a birefringent optical response is artificially induced, so that an extraordinary optical axis arises in the direction perpendicular to the surfaces plane. Consequently, the optical response of the HMM in the plane parallel to its surface results dramatically different from the one detected in its bulk. If the number of bilayers is large enough and their thickness is deeply subwavelength, it is possible to homogenize the optical response of the HMM. In this case, the uniaxial overall optical response can be described using an *effective dielectric permittivity*, whose components parallel and perpendicular to the HMM's surface plane can be calculated by applying simple boundary conditions [21].

The constitutive relations remain valid independently on the polarization of the incident electric field and even for the effective medium so that they can be used for the calculation of both the parallel and perpendicular effective permittivity components:

$$D_j = \tilde{\varepsilon}_j E_j \tag{5.1}$$

where j stands for the considered material that can be the metal or the dielectric. It is important noticing that Eq. (5.1) is still valid even for the overall HMM, so that it can be assumed that $D_{eff} = \varepsilon_{eff} E_{eff}$, where ε_{eff} can be equal to $\varepsilon \|$ or $\varepsilon_{\perp}$, accordingly to which parameter is being calculated.

Let's now consider the case in which the electric field is polarized along the HMM's surface plane. This procedure leads to the calculation of $\varepsilon \|$. In this case, the electric field has to be continuous at the interfaces between the metal and the dielectric. This means that:

$$E_m = E_d = E_{eff} \tag{5.2}$$

where the subscripts *m, d* and *eff* stands for *metal*, *dielectric* and *effective,* respectively. The superscript "~" in Eq. (5.1) indicates that the dielectric permittivities are complex quantities. Indeed, the effective parameters are also valid for the imaginary components and are intrinsically Kramers-Krönig consistent.

The effective electric flux can be expressed as the sum of the flux density in the metal and the dielectric, averaged to each respective fill fraction $f_j = (t_{j/}(t_{tot}))$, being t_j the thickness of the jth material ant t_{tot} the total thickness of a single metal/dielectric bilayer:

$$D_{eff,\,||} = \tilde{\varepsilon}_{||} E_{eff} = f_m D_m + f_d D_d \tag{5.3}$$

by replacing Eq. (5.2) in Eq. (5.3), we obtain the expression for $\tilde{\varepsilon}_{||}$.

$$\tilde{\varepsilon}_{||} = \frac{\tilde{\varepsilon}_d t_d + \tilde{\varepsilon}_m t_m}{t_d + t_m} \tag{5.4}$$

In order to calculate $\tilde{\varepsilon}_{\perp}$ we start from the consideration that, when the electromagnetic field is polarized perpendicularly to the HMM's surface plane, the electric flux D_j is continuous at the interfaces:

$$D_{eff,\perp} = \tilde{\varepsilon}_{\perp} E_{eff} = D_m + D_d \tag{5.5}$$

In this case, the effective electric field can be expressed as the fill-fraction averaged electric field in every single component, so that

$$E_{eff} = f_m E_m + f_d E_d \tag{5.6}$$

and, by replacing Eq. (5.6) in Eq. (5.5), the expression $\tilde{\varepsilon}_{\perp}$ can be immediately found:

$$\tilde{\varepsilon}_{\perp} = \frac{\tilde{\varepsilon}_d \tilde{\varepsilon}_m (t_m + t_d)}{\tilde{\varepsilon}_m t_d + \tilde{\varepsilon}_d t_m} \tag{5.7}$$

Equations (5.4) and (5.7) represent the two components of the *effective local permittivities* for a multilayered HMM.

Provided that the optical response can be modeled with sufficient precision via the local $\tilde{\varepsilon}_{||}$ and $\tilde{\varepsilon}_{\perp}$, it is possible to calculate the exact dispersion relation for a multilayered HMM. Such a quantity provides information about the light propagation through the metamaterial and allows explaining a plethora of noticeable properties, from super-collimation to lifetime engineering in the proximity of HMMs.

In general, the dispersion relation describing the wavelength dependence of the wavevector can be calculated by starting from Maxwell equations in the harmonic regime:

$$\bar{k} \times \overline{E} = \omega\mu_0\overline{H} \tag{5.8}$$

$$\bar{k} \times \bar{H} = -\omega\overleftrightarrow{\varepsilon}\bar{E} \tag{5.9}$$

where $\bar{k}$ is the wavevector, ω is the frequency, μ_O is the vacuum permittivity and $\overleftrightarrow{\varepsilon}$ is the complex dielectric permittivity tensor. In the most general case where the dielectric permittivity of the material changes in all the three spatial directions, $\overleftrightarrow{\varepsilon}$ assumes the form of a diagonal matrix with components ε_{xx}, ε_{yy} and ε_{zz}. Multiplying both the terms on the left and right side of Eq. (5.8) by $\bar{k}\times$, it turns out that:

$$\bar{k} \times \bar{k} \times \bar{E} = \omega\mu_0\bar{k} \times \bar{H} \tag{5.10}$$

Remembering that $\omega = k_O\, c$, being c the speed of light in vacuum and that $c = (\varepsilon_O\mu_O)^{-1/2}$, replacing Eq. (5.9) in (5.10) leads to:

$$\bar{k} \times \bar{k} \times \bar{E} - k_0^2\overleftrightarrow{\varepsilon}\bar{E} = 0 \tag{5.11}$$

Equation (5.11) constitutes the eigenvalue equation for the electric field. A general uniaxial crystal with the extraordinary optical axis along the z-direction is characterized by a dielectric permittivity tensor equal to:

$$\overleftrightarrow{\varepsilon} = \begin{bmatrix} \varepsilon_{xx} & 0 & 0 \\ 0 & \varepsilon_{xx} & 0 \\ 0 & 0 & \varepsilon_{zz} \end{bmatrix} \tag{5.12}$$

In this case, Eq. (5.11) can be manipulated to obtain the general dispersion relation for a uniaxial crystal:

$$\frac{k_x^2 + k_y^2}{\varepsilon_{zz}} + \frac{k_z^2}{\varepsilon_{xx}} = \frac{\omega^2}{c^2}; \tag{5.13}$$

Since in an HMM $\varepsilon_x = \varepsilon_y = \varepsilon_{||}$ and $\varepsilon_z = \varepsilon_\perp$, Eq. (5.13) becomes:

$$\frac{k_x^2 + k_y^2}{\varepsilon_\perp} + \frac{k_z^2}{\varepsilon_{||}} = \frac{\omega^2}{c^2}; \tag{5.14}$$

That is the dispersion relation for an HMM modeled via the effective parameters $\varepsilon_{||}$ and $\varepsilon_{\perp}$. Since the sign of $\tilde{\varepsilon}_{||}$ and $\tilde{\varepsilon}_{\perp}$ can be either positive of negative, four cases can be distinguished:

1. $\varepsilon_{\perp} > 0$ and $\varepsilon_{||} > 0$: The HMM behaves as an *effective anisotropic dielectric.*
2. $\varepsilon_{\perp} < 0$ and $\varepsilon_{||} > 0$: The HMM manifests the so-called *type I* anisotropy, behaving as a metal in the bulk and as a dielectric in the plane.
3. $\varepsilon_{\perp} > 0$ and $\varepsilon_{||} < 0$: The HMM manifests the so-called *type II* anisotropy, behaving as a dielectric in the bulk and as a metal in the plane.
4. $\varepsilon_{\perp} < 0$ and $\varepsilon_{||} < 0$: The HMM behaves as an *effective anisotropic metal.*

By fixing the frequency and varying the wavevectors, it is possible to draw the *equifrequency contours* of Eq. (5.14). In the case of a conventional anisotropic dielectric like a liquid crystal, the resulting *equifrequency contour* graph is an ellipsoid in the wavevector space. Therefore, for a precise frequency, the permitted wavevectors result confined in an ellipsoid whose larger and smaller semi-axes are determined by the extraordinary and ordinary refractive index values. The case illustrated in point 1, in which the HMM behaves as an *effective dielectric*, resembles the features just described. As mentioned before, in an HMM the in-plane and perpendicular effective permittivities can be of opposite sign. If this happens, Eq. (5.14) assumes the shape of an open hyperboloid for the *type I* anisotropy and of a closed hyperboloid for the *type II*. The case of the anisotropic metal has no solution in the real wavevector plane but leads to a dispersion similar to that calculated in the case of the effective dielectric, in the complex wavevector space. Figure 5.1a illustrates a sketch of a typical multilayered HMM, while Fig. 5.1b shows the bidimensional conical section in the $K_x - K_z$ plane of the equifrequency contours for the hyperbolic anisotropies. The *type I* regime is characterized by an open hyperboloid while for the *type II* the hyperboloid is closed. In these two cases, there are ideally no upper limits in the values of K_x and K_z, indicating that very high wavevectors are allowed in these materials. Modes propagating in the hyperbolic regimes with such huge wavevectors are usually called *high-k* modes. *High-k* modes allow, for example, an additional near-field radiative channel for fluorophores placed in proximity of the HMM and are, therefore, responsible for significant modification of the PDOS [1]. Engineering of high-k modes has been extensively used as a way to achieve broadband Purcell effect. Moreover, they proved to be the key element for reaching extreme resolution bio-sensing [7].

As already discussed, HMMs allow for the design of unprecedented anisotropy, whose features can be tailored by carefully choosing their fundamental components and by simply acting on their geometrical parameters. Particularly interesting is the extreme case in which the parallel dielectric permittivity is designed to be very close to zero and the perpendicular one to be very high. Such a regime represents an anisotropic limit and is called *epsilon-near-zero-and-pole* (ENZP) and is usually verified for only a single wavelength in a given HMM. In Fig. 5.1b, the equifrequency contour for an Epsilon-Near-Zero and Pole (ENZP) HMM is shown (black curves). For such a system, the hyperboloid is completely flattened in two quasi-parallel planes facing each other. Very high wavevectors are allowed in the K_x plane, but the

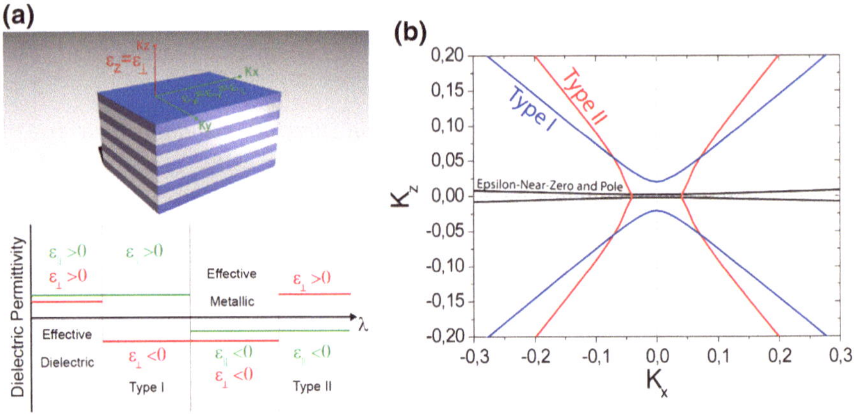

Fig. 5.1 **a** Sketch of a multilayered hyperbolic metamaterial with referenced all the three homogenized dielectric permittivities. On the bottom, a scheme of the four possible combinations with relative anisotropy type. **b** On the right, the bidimensional conical section in the x-z plane representing the equifrequency contour for the (red) *type II*, *type I* (blue) and the particular case of an Epsilon-Near-Zero and Pole (black) HMM, treated in this chapter

wavevector K_z remains ideally constant at a value very close to zero. Even though in reality K_z slightly changes with respect to K_x, its evolution is so slow that very few values are allowed for K_z in real HMMs, in which the lateral XY size is limited. For these real systems, K_z always remains very close to zero. Such a feature is crucial for achieving properties described few after, among which one noticeable is the supercollimation. In the ENZP regime, extreme light localization and confinement can be achieved, opening the way to nano-resolution imaging and ultra-collimation of light of plasmonic solitons propagating in the bulk of the metamaterial. In this chapter, it will be explained how to engineer the ENZP condition to occur in a specific HMM and the exceptional light manipulation capabilities, allowed in such a regime, going from super-collimation to nano-imaging, will be illustrated.

5.3 Design of the Epsilon-Near-Zero-and-Pole Condition

When engineering the optical properties of an HMM in the framework of the EMT, the imaginary part of the dielectric permittivity of the metallic and dielectric components cannot be neglected since too high losses can prevent the occurrence of the ENZP regime. Naturally, the permittivities expressed in Eqs. (5.4) and (5.7) are complex numbers, and their imaginary parts play a crucial role in the optical response of the HMM. However, when common oxides like SiO_2, TiO_2, and Al_2O_3 are used in combination with conventional plasmonic metals like Au and Ag the imaginary part of Eqs. (5.4) and (5.7) can be neglected. Such a simplification promotes an intuitive understanding of the effective optical response of the HMM. For instance, it can be readily understood that it is challenging to obtain a *type I* anisotropy in the visible range with multilayered HMMs. However, a quick inspection shows that Eq. (5.4) can manifest a zero exactly where Eq. (5.7) manifests a pole. Such a condition is known as *Epsilon-Near-Zero-and-Pole* (ENZP). In order to achieve an ENZP singularity, two conditions have to be verified simultaneously:

$$t_m = t_d \tag{5.15}$$

$$\varepsilon_m(\lambda) = -\varepsilon_d(\lambda) \tag{5.16}$$

The first condition is easily assessable by depositing metallic and dielectric layers of equal thickness. The second condition is far more stringent, and, for a given metal/dielectric pair, it can be verified only at one wavelength. Once the condition posed by Eq. (5.15) is satisfied by designing 50% fill fraction HMM, Eq. (5.16) can be simply graphically solved, providing a quick and easy design tool for ENZP wavelength. Figure 5.2a shows the crossing points between the absolute value of the real dielectric permittivity of the two most commonly used plasmonic metals (Ag, dashed blue curve and Au, orange dashed curve), and that of the mainly used oxides. Higher permittivity dielectrics push the ENZP wavelength towards the infrared spectral region. As shown in Fig. 5.2a, Ag/SiO_2 pair satisfy Eq. (5.16) at 365 nm, where the absolute value of the real dielectric permittivity of Ag crosses the one of SiO_2. This wavelength constitutes the ENZP wavelength for an Ag/SiO_2 HMM. At wavelengths shorter than the ENZP, the HMM manifests a *type I* anisotropy, while for higher wavelengths the system becomes a *type II* HMM. Therefore, the ENZP wavelength constitutes the point of inversion of two extreme anisotropies coexisting in the same HMM. On the other hand, the absolute value of real dielectric permittivity of Au crosses that of SiO_2 at 480 nm. Figure 5.2b shows the real parts of the effective parallel and perpendicular effective dielectric permittivity calculated using Eqs. (5.4) and (5.7) respectively, for two different HMMs consisting in Ag/SiO_2 (blue curves) and Au/SiO_2 (red curves) multilayers, with 50% metal/dielectric fill fraction. As predicted, at 365 nm the real parallel permittivity of the Ag/SiO_2 HMM manifests a zero and the perpendicular one a pole, recognizable as a pronounced Lorentz sigmoid with a flex at 365 nm, representing the zero-crossing point of $\varepsilon_\perp$. This is the signature of the occurrence of the ENZP condition for this system.

On the contrary, even though ε’|| of the Au/SiO_2 HMM shows a zero at 480 nm, $\varepsilon_\perp$ does not show a pole, preventing the occurrence of the ENZP regime in such an HMM. This is mainly because the ENZP condition for Au/SiO_2 HMM falls in Au interband transition range, where losses are too high to let this condition arise. This effect points out the role played by losses in designing metamaterials in the visible spectrum. The high quest for low-loss metals brought to investigate novel compounds, like TiN that revealed advantageous with respect of the classically used noble metals under many aspects [26].

One emerging class of metals holding very performing optical parameters are the alkali metals. Figure 5.3 shows the real and imaginary effective dielectric permit-

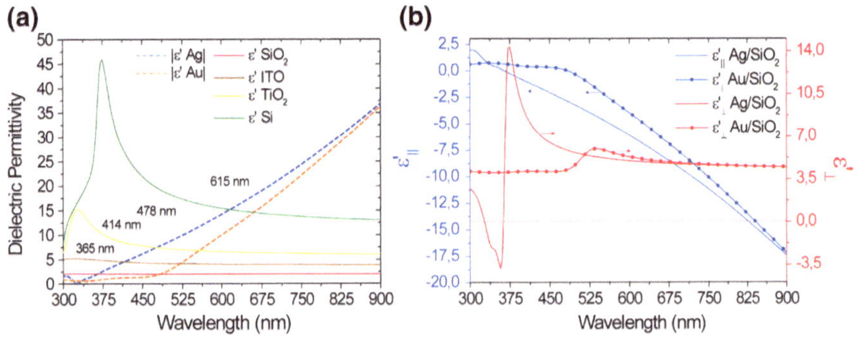

Fig. 5.2 **a** Crossing points between the absolute value of the real dielectric permittivity of Ag and Au, in comparison with the most commonly used dielectrics. The crossing points represent the possible ENZP wavelengths for the specific metal/dielectric pair. **b** Calculated effective parallel and perpendicular real dielectric permittivity for a Ag/SiO_2 and an Au/SiO_2 HMM

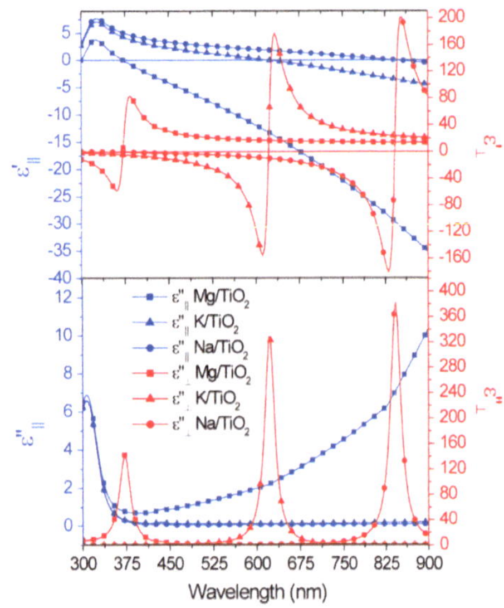

Fig. 5.3 Effective parallel and perpendicular dielectric permittivity for 50% fill fraction HMMs made of TiO_2 as a dielectric and Mg (squares), K (triangles) and Na (circles), showing ENZP wavelengths in the UV, VIS and NIR range

tivities of three different ENZP HMMs, involving Mg, K, and Na as metals. Alkali metals have the noticeable advantage of possessing a relatively small negative real dielectric permittivity, together with very low losses in the visible range, opening to the possibility of positioning the ENZP wavelength in usually inaccessible bands. For example, Mg shows negligible losses in the UV range and, thanks to its easiness of hydrogenation, opens to interesting sensing possibilities. On the other hand, due to their very small negative dielectric permittivity, K and Na allow placing the ENZP wavelength deep in the visible range and in the NIR, respectively. Unfortunately, dealing with alkali metals is not straightforward since their high reactivity in the air requires special passivation techniques.

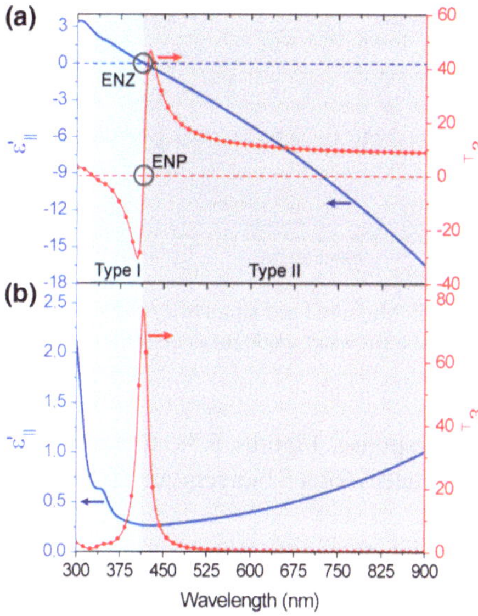

Fig. 5.4 **a** Real and **b** imaginary planar and perpendicular effective dielectric permittivities for a 50% fill fraction Ag/ITO HMM, manifesting an ENZP condition at 414 nm. Adapted from Ref. [8]

5.4 Far-Field Analysis and Scattering Parameters of Ag/ITO ENZP HMM

When the constraints required in Eqs. (5.15) and (5.16) are fulfilled, an ENZP wavelength in the effective dielectric permittivity of the HMM manifests. Figure 5.4 shows the effective permittivity of an ENZP HMM made of Ag/ITO pairs with 50% fill fraction. It can be noticed that: (i) before 327 nm (being the volume plasmon of Ag), the HMM behaves as an effective dielectric, since both $\varepsilon_{||}$ and $\varepsilon_{\perp}$ are greater than zero. (ii) between 327 nm and 414 nm $\varepsilon_{\perp} < 0$ while $\varepsilon_{||} > 0$. This is the *type I* anisotropy, in which the HMM behaves as a metal in the *Z*-direction (bulk) and as a dielectric in *XY* the plane. Above 414 nm the HMM manifests a *type II* anisotropy, since $\varepsilon_{\perp} > 0$ while $\varepsilon_{||} < 0$. Figure 5.4a well highlights the point of inversion between the two extreme anisotropies coexisting in the same metamaterial. It is important noticing here that the z-axis (bulk) propagation across the *type I/type II* transition is characterized by intrinsically high losses, that manifest as a sharp Lorentzian peak in the imaginary part of the effective perpendicular permittivity. This means that common far field measures like the evaluation of the classic scattering parameters (reflectance and transmittance) cannot show any significant feature at the ENZP wavelength. As will be shown further in the chapter, the presence of the dielectric singularity discloses in near field investigations, where the evanescent nature of the light traversing the HMM can be appreciated. Here, the exceptional light collimation and confinement properties typical of the dielectric singularity become evident.

Due to its intrinsic metallic nature, the *type II* region is very reflective while, on the contrary, being principally dielectric, the *type I* manifests a more transmis-

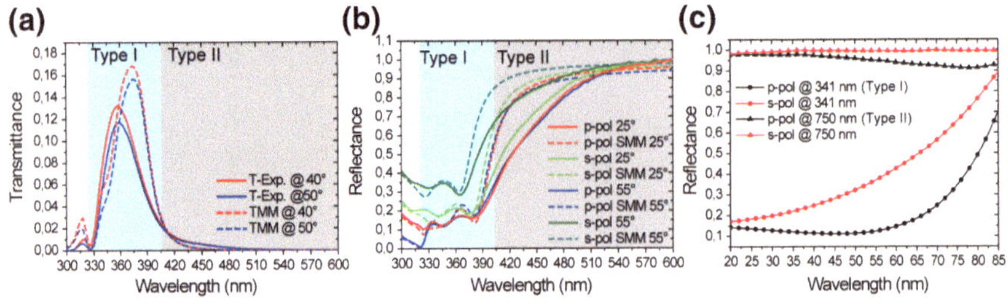

Fig. 5.5 Measured and simulated *p*- and *s*- polarized **a** Transmittance and **b** Reflectance, together with **c** the Brewster angle measure for the considered ENZP HMM. Adapted from Ref. [8]

sive response. Figures 5.5a, b show the scattering parameters (transmittance and reflectance) for a 5 bilayers Ag/ITO HMM, whose layers have an equal thickness of 20 nm.

Many interesting features can be individuated after an in-depth inspection of the transmittance and reflectance spectra shown in Fig. 5.5a, b. The dip present in the two graphs at around 327 nm represents the volume plasmon of silver, a feature that occurs naturally in this metal. The classic free electrons Drude model for Ag can predict the occurrence of this mode, framing it as a volume plasmon characterized by a low-loss zero crossing of the dielectric permittivity [27]. This mode is also called *Ferrell-Berreman* mode. Moreover, both *p*- and *s*-polarized reflectance spectra manifest several dips in the *type I* region. These modes have been recently identified as "Fabry-Pérot like" cavity modes called *Bulk Plasmon Polaritons* (BPPs), arising from the hybridization of *Gap Plasmon Polaritons* originating in every single metal/dielectric/metal cavity composing the HMM. It has also been suggested that these modes can be seen as an "engineered" version of the natural *Ferrell-Berreman* mode occurring in Ag at 327 nm, framing them as artificial Epsilon-Near-Zero modes [29]. However, the simple framework of the effective medium theory in which usually HMMs are modeled, is not able to catch their ENZ nature, and more sophisticated theories have to be developed. One way to confirm the reflective (transmissive) characteristics of the *type II* (*type I*) regime, is evaluating the absence (presence) of the Brewster angle for wavelengths falling in this band. Figure 5.5c shows the ellipsometrically measured *p*- and *s*-polarized reflectance for two different wavelengths, one falling in the *type I* band (341 nm) and another one in the *type II* (750 nm). As expected, for the *type I* wavelength a Brewster angle (pseudo) is detected in the *p*-polarized reflectance around 50°. No Brewster angle is detected at 750 nm, where the HMM is mainly metallic.

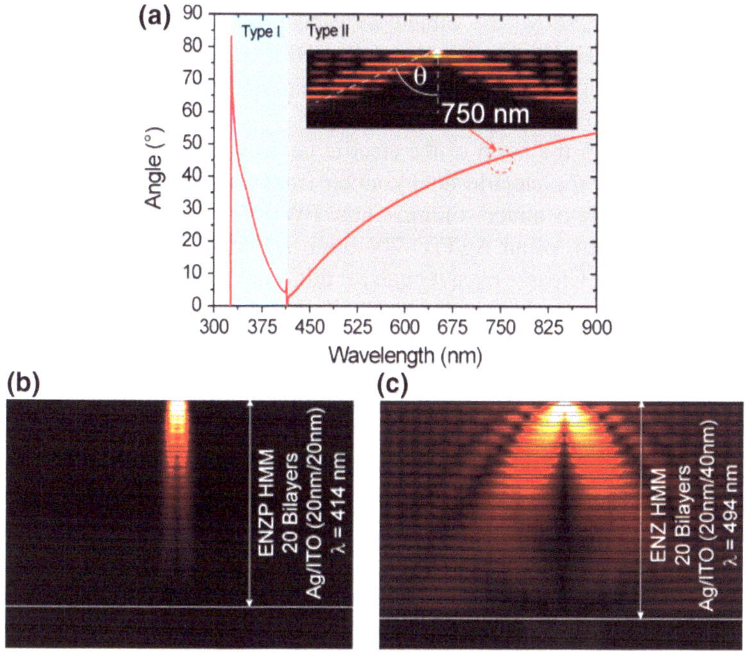

Fig. 5.6 **a** Resonance cone calculation for the considered ENZP HMM. **b** Supercollimation at the ENZP wavelength and **c** propagation through a simple ENZ HMM. Adapted from Ref. [8]

5.5 Light Propagation at the ENZP Wavelength and Supercollimation Effect

Light propagation inside an HMM can be rigorously described by means of the dyadic Green's function. Such an analysis leads to the conclusion that light emitted in the direction of the extraordinary axis from a localized source placed on the top of a HMM, propagates within the HMM as a so-called *resonance cone*. Such a concept is expressed in Fig. 5.6a, where a 50% fill fraction Ag/ITO HMM is excited with a vertically oriented point dipole at 750 nm, falling in the hyperbolic region. The resonance cone is visible as two lobes propagating through the HMM, separated by a semi-angle θ. The propagation in resonance-cones is the spatial counterpart of the hyperbolic dispersion occurring in the isofrequency contours of the dispersion relation. The aperture θ of the resonance cone can be calculated as follows [1]:

$$|\theta| < \sqrt{-\frac{\varepsilon_{||}}{\varepsilon_{\perp}}} \tag{5.17}$$

If $\varepsilon_{||} \to 0$ or $\varepsilon_{\perp} \to \infty$ or, even better, $\varepsilon_{||} \to 0 \, \Lambda \, \varepsilon_{\perp} \to \infty$, then θ goes to zero and light is expected to propagate through the HMM as a well confined solitary wave (soliton). Figure 5.6a shows the value θ as a function of the wavelength, in a 50% fill fraction Ag/ITO HMM. Noticeably, at the ENZP wavelength equal to 414 nm, $\theta = 0$. However, a simple ENZ condition as in the case in which only $\varepsilon_{||} \rightarrow 0$ is not sufficient to reach this extreme collimation effect.

Figure 5.6b shows the norm of the electric field for a 50% fill fraction Ag/ITO HMM. In this case, the electric field coming from a vertically oriented dipole is collimated in a highly confined soliton, whose FWHM is about 50 nm. In Fig. 5.6c, the electric field norm of an Ag/ITO 20/40 nm HMM at the ENZ wavelength of 494 nm is presented. Clearly, even though for this latter HMM $\varepsilon_{||} \rightarrow 0$, no supercollimation effect is present. Indeed, for an ENZP HMM it is always verified that $K_z = 0$. Therefore, light propagating inside the HMM in the ENZP regime experiences ideally no phase change and traverses the HMM as its front and back layer are in immediate contact. Obviously, losses present in the metal prevent such ideal condition from occurring and attenuation is present. One way of overcoming this drawback is compensating the losses by employing a suitably dimensioned fluorophore as a dielectric. Such a condition will be treated in Chap. 6. The supercollimation effect can be investigated by exciting the ENZP HMM with strongly localized light sources.

Figure 5.7a shows the confocal analysis of the transmission through a 50% fill fraction HMM made of 20/20 nm Ag/ITO bilayers. In this experiment, light coming from a pulsed laser was focused on top of the HMM by means of a 100× objective (0.75 NA) and collected via a 50× objective (0.5 NA). No transmitted beam is detected when analyzing the sample at 530 nm, in the *type II* region, confirming the highly reflective behavior of the HMM in this regime. Interestingly, even though it is known that shorter wavelengths are focused in smaller diameters (thus usually allowing higher resolutions), the Full Width at Half-Maximum (FWHM) of the transmitted beam at 355 nm (*type I* regime) is much broader than the one measured at 414 nm, being the ENZP wavelength for such HMM. This confirms the supercollimation effect occurring at the ENZP wavelength. Such an effect is much more pronounced as much the excitation source approaches the periodicity of the HMM, ideally approximating the point dipole. An interesting possibility to approach such an experimental scenario is illuminating the ENZP HMM via Scanning Near field Optical Microscopy (SNOM). Such a technique makes use of nano-metric hollow tips whose hole diameter is in the range of few tenths of nanometers.

In order to numerically simulate the typical SNOM excitation by means of a Finite Element Method approach (COMSOL Multiphysics), it possible to excite the ENZP HMM from a nanometric slit placed at the top of the HMM. Figure 5.7b shows the case of the same HMM as before, illuminated from a 40 nm slit (equal to the period of the HMM) placed on the top of the HMM at 414 nm (ENZP wavelength). Clearly, light propagates inside the HMM as a well-confined soliton, with a FWHM comparable to the period of the HMM. Noticeably, as confirmed in the inset of Fig. 5.7b, a propagation length of the supercollimated beam of more than 100 Rayleigh lengths is found.

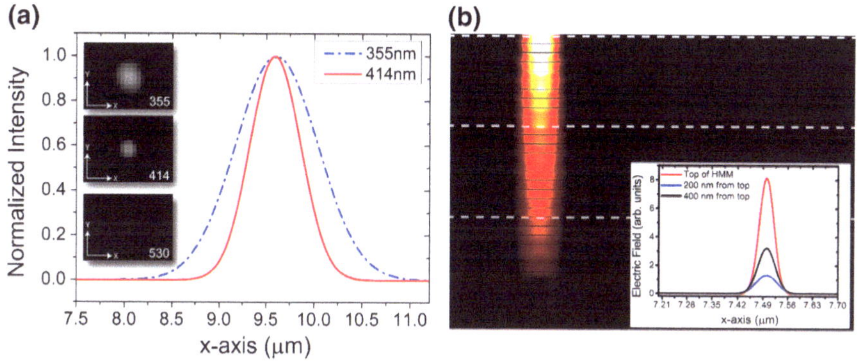

Fig. 5.7 Confocal analysis of the ENZP HMM. Light impinging on the HMM at the ENZP wavelength shows a much smaller FWHM with respect to a shorter wavelength in the *type I* regime. **b** Supercollimation through the HMM of light coming from a 40 nm width excitation slit. Propagation in the HMM lasts for more than 100 Rayleigh lengths. Adapted from Ref. [8]

5.6 ENZP Perfect Lens

The supercollimation effect described in the previous section has the noticeable consequence of allowing the possibility to achieve extremely high-resolution imaging. In 2003, Ramakrishna et al. described the mechanism lying behind the super-resolution achievable in a similar system [28]. It is convenient to decompose the electromagnetic radiation scattered or emitted in the *x*-*y* plane by a 2D object, into its spatial (k_x and k_y) and polarization (σ) Fourier components:

$$E(x, y, z; t) = \sum_{k_x, k_y, \sigma} E_\sigma(k_x, k_y, k_z) \exp[i(k_x x + k_y y + k_z z - \omega t)] \qquad (5.18)$$

Since k_x, k_y, and k_z are complex quantities, a further decomposition in real and imaginary components is possible. The real components describe the coarse details of the image. They are associated with propagating waves and constitute the low-frequency content of the Fourier expansion. The finest details are, instead, included in the evanescent components, constituting the high-frequency content. The dependence on z of the amplitude of each Fourier component makes the image distorted in the *x*-*y* plane while propagating along z. Common dielectric lenses can account only for phase correction in the low-frequency details so that the high-frequency component evanescently decays, letting the nanometric high-frequency details to be lost. In order to deal with the evanescent, high-frequency field component, plasmonic materials are needed. Even a simple single layer of silver can be used to improve the resolution in the near field [29]. However, the single slab plasmonic lens results extremely sensitive to the absorption of the plasmonic layer. On the other hand, as demonstrated in Fig. 5.7b, light propagating in a multilayered ENZP HMM has the special characteristic of evolving symmetrically in the metallic and dielectric slabs, causing the generation of the supercollimated soliton. Equation (5.14) reveals also that in an ENZP HMM, the component k_z is always equal to zero. Therefore, the phase-front of a wave propagating along the z-direction in such a material experiences no changes. The xy high-frequency nanometric Fourier components of the electric field, scattered or radiated from a 2D object placed at the top of the HMM, propagate as straight solitons within the HMM, appearing unvaried at the exit. The great advantage of this effect is that, in the limit of the near field, the resolution achievable in such a system is no more related to the classic diffraction limit but, rather, is equal to [28],

$$\Delta = 2\pi \varepsilon_m^{''}(\lambda) d \tag{5.19}$$

being $\varepsilon_m^{''}$ the imaginary dielectric permittivity of the metal and d the thickness of the single slab.

Figure 5.8a–d show the Finite Element-based simulation of the imaging of three nanometric objects placed on the top of an ENZP HMM (by COMSOL Multiphysics). In particular, a monochromatic plane wave coming from an excitation slit of 500 nm, illuminates three polymeric element ($n = 1.45$) positioned on the top of a simple glass slab (Fig. 5.8a, c) of thickness equal to the complete HMM (Fig. 5.8b, d). The three elements are polymeric rectangles (height = 100 nm and width = 40 nm) separated by 40 nm from each other. The analysis is conducted at two different wavelengths, 350 nm (*type I* range) and 414 nm (supercollimation wavelength). Clearly, in the cases of glass slab as well as in that of the HMM operating in the *type I* region, no supercollimation is present. On the contrary, when the HMM is illuminated at its ENZP wavelength, the supercollimation effect induces all the components of the image, coming from the three objects placed on the top of the HMM, to be perfectly preserved and accessible at the exit layer. Figure 5.8e, f show the electric field norm profile collected at the exit of the HMM. At the ENZP wavelength, the three objects are perfectly visible as three minima, whose peaks are positioned 80 nm apart from each other, perfectly reproducing the profile of the three polymeric elements on the HMM.

It is worth noticing that, at the ENZP wavelength, the radiation propagates within the HMM almost unvaried, just as the top and bottom layers were in close contact. In order to experimentally demonstrate the possibility of reaching nanometric resolution with such systems, in Fig. 5.8g–i we propose an experimental proof-of-concept of a real system. A single ITO/Ag bilayer, with a fill fraction of 50%, was fabricated, in order to keep the transmissivity of the device suitably high. A sketch of the proposed experiment is depicted in Fig. 5.8g. A simple 1D grating consisting of subwavelength strips of 250 nm width and periodicity of 500 nm, was used as the structure to be imaged at the exit of the HMM. A confocal analysis has been carried out, trying to resolve the grating at the exit of the HMM, illuminating at the same wavelengths proposed in the simulations. Noticeably, even though in principle confocal imaging shows higher resolution when conducted with shorter wavelength, we demonstrate that the grating is only visible when the experiment is carried out at the ENZP wavelength (414 nm).

5.7 Three Materials ENZP HMMs

One of the main drawback by which HMMs are affected is their lack of tunability. Do to the nanometric size of the fundamental component imposed by the homogenization rules, many of the most common tunable dielectrics, like for example liquid

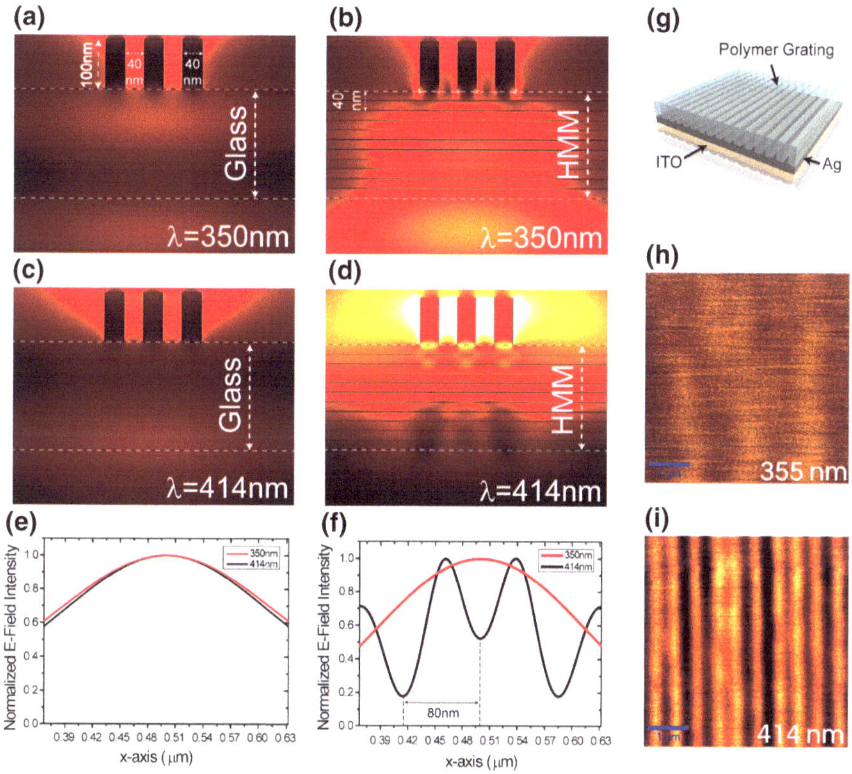

Fig. 5.8 FEM simulations of a plane wave exciting the model from a slit of 500 nm above three PMMA nanometric elements (100 nm width, 10 nm separation distance) through a glass slab of thickness 280 nm, at two wavelengths (355 and 414 nm). **b**, **d** represent the FEM simulations with the glass slab replaced by the HMM made of 5 bilayers (Ag-ITO 20–20 nm). Analysis of the transversal intensity profiles made at the exit of the glass/HMM (**e**, **f**), shows that the system is able to resolve the nanometric structure with a resolution comparable to the step of the metamaterial. **g** Sketch of a proof of concept experiment made on a single Ag/ITO bilayer with a grating on the top (250 nm stripes and 250 nm gaps). **h**, **i** Confocal analysis of the periodic structure at $\lambda = 355$ nm (**h**) and 414 nm (**i**). Adapted from Ref. [8]

crystals, cannot be included in these systems. Some attempts of embedding tunable dielectrics have been provided so far, with special attention to electrically tunable conductive oxides (ITO is one promising candidate in this direction), chalcogenides and graphene [30]. Unfortunately, these options often require significant technological and nano-fabrication efforts to be realized and the tuning ranges are no more than a handful of nano-meters. In Sect. 5.5.3 it has been shown that for a HMM composed by a metal/dielectric bilayer there is only one single wavelength for which the ENZP condition is verified. This limit can be overcame by adopting a new configuration in which the fundamental unit of the ENZP HMM consists in a metal/high-k dielectric/low-k dielectric tri-layer. Such a configuration allows extending the design band of ENZP HMMs in the entire visible range [31].

In the special case of a three-materials HMM, the aforementioned expressions for $\tilde{\varepsilon}_{||}$ and $\tilde{\varepsilon}_{\perp}$ have to be modified. Nonetheless, the same considerations that brought to the expression of the homogenized dielectric permittivity for a two-materials HMM are still valid and Eqs. (5.4) and (5.7) can be immediately extended to the case of three materials HMM:

$$\tilde{\varepsilon}_{||} = \frac{\tilde{\varepsilon}_{d1} t_{d1} + \tilde{\varepsilon}_{d2} t_{d2} + \tilde{\varepsilon}_m t_m}{t_{d1} + t_{d2} + t_m} \tag{5.20}$$

$$\tilde{\varepsilon}_{\perp} = \frac{\tilde{\varepsilon}_{d1}\,\tilde{\varepsilon}_{d2}\,\tilde{\varepsilon}_m\,(t_m + t_{d1} + t_{d2})}{t_{d1}(\tilde{\varepsilon}_{d2}\,\tilde{\varepsilon}_m) + t_{d2}(\tilde{\varepsilon}_{d1}\,\tilde{\varepsilon}_m) + t_m(\tilde{\varepsilon}_{d1}\,\tilde{\varepsilon}_{d2})} \tag{5.21}$$

Here the subscripts d_1, d_2 and m respectively indicate the first and second dielectric and the metal, while t_{d1}, t_{d2} and t_m are their thicknesses. Once more, in order to verify the conditions $\varepsilon_{//} \to 0$ and $\varepsilon_{\perp} \to \infty$ at the same wavelength, it is convenient fixing the thickness of the metal at the desired value and impose the numerator of Eq. (5.20) and the denominator of Eq. (5.21) go to zero. This produces the following system:

$$\begin{cases} t_{d_1} = -\frac{\tilde{\varepsilon}_{d2} t_{d2} + \tilde{\varepsilon}_m t_m}{\tilde{\varepsilon}_{d1}} \\ t_{d_2} = -t_m \frac{\left(\tilde{\varepsilon}_{d_1}^2 - \tilde{\varepsilon}_m^2\right)}{\left(\tilde{\varepsilon}_{d_1}^2 - \tilde{\varepsilon}_{d_2}^2\right)} \frac{\tilde{\varepsilon}_{d2}}{\tilde{\varepsilon}_m} \end{cases} \tag{5.22}$$

Since we are only interested in positive thicknesses of the two dielectrics, the conditions $t_{d_1} > 0$ and $t_{d_2} > 0$ can be imposed. The dielectric permittivity of the metal is taken negative in the whole range of interest, while that of the dielectric is positive. Therefore, in order to achieve positive thicknesses of the dielectrics, it has to be that $\tilde{\varepsilon}_{d_1} < |\tilde{\varepsilon}_m|$ and $\tilde{\varepsilon}_{d2} > |\tilde{\varepsilon}_m|$. Therefore, merging these two conditions leads to the design equation for a three-materials ENZP HMM:

$$\tilde{\varepsilon}_{d_2} > |\tilde{\varepsilon}_m| > \tilde{\varepsilon}_{d_1} \tag{5.23}$$

In order to extend the validity range of Eq. (5.23), a very low and very high index dielectric have to be used. It is worth noticing that the two points in which $\tilde{\varepsilon}_m$ intersects one of the two dielectric's curves do not constitute acceptable solutions, because here a thickness equal to zero for one of the two dielectrics is required, degenerating in the case of a two-material ENZP HMM. Now, we will try to engineer ENZP HMMs reasonably far from these two points, in order not to be forced at using too thin dielectrics.

Figure 5.9a shows all the possible triplets coming from the combination of silver and few of the most common dielectrics (Si, SiO_2 and TiO_2). As shown in Fig. 5.9b, the most convenient triplet is the one constituted by Si (high index dielectric) and SiO_2 (low index), being its natural oxide. Such a combination ensures a design range of 255 nm. Figure 5.9b reports the thicknesses of the two dielectrics to be used for obtaining an ENZP HMM in the entire tuning range, calculated via Eq. (5.23). At 365 nm, the trilayer system becomes an Ag/SiO2 HMM, while at 620 nm the system assumes the characteristics of an Ag/Si bilayer HMM. With the aim of demonstrating the tunability of the trilayer HMM, two different configurations are engineered, one showing the ENZP condition at $\lambda_{tr,1} = 450$ nm and the other at $\lambda_{tr,2} = 550$ nm. Thickness of the silver layer is fixed at $t_{Ag} = 20$ nm for both the configurations. By solving the system in Eq. (5.23) for $\lambda = 450$ nm, the two dielectric layers thicknesses are: $t_{SiO2} = 6.5$ nm and $t_{Si} = 5$ nm, while for $\lambda = 550$ nm it results that $t_{SiO2} = 2$ nm and $t_{Si} = 13.5$ nm. In Fig. 5.10, the EMT design of the two samples is reported. As showed in Fig. 5.10a, the designed ENZP HMM behaves as a highly anisotropic dielectric in the deep UV range until 327 nm, where it undergoes its first effective dielectric/*type I* phase transition. At 450 nm the ENZP condition is satisfied and a dielectric singularity occurs.

A dramatic topological transition from *type I* to *type II* anisotropy occurs, manifesting as a strong Lorentzian sigmoid in the real effective perpendicular permittivity and as a sharp peak in the respective imaginary part. From this wavelength on, the HMM behaves as *a type II* HMM, with marked metallic characteristics. The same happens for the second ENZP HMM at 550 nm. All the transmissive features already described in the case of the two-materials ENZP HMM (see Fig. 5.5a, b) are present also in the cases of the three-materials HMM, as well as the presence of the Ferrel-Berreman mode, as shown in Fig. 5.5a, b.

Figure 5.11 shows the transmission and reflection spectra of the two HMMs, showing the typical transmissive features in the respective *type I* range and very reflective behavior in the *type II*. Since the dispersion relation for a three-materials ENZP HMM remains unvaried with respect to that of the classic two-material version, no differences in the light propagation features are present in the case of the new configuration. The supercollimation effect, as well as the resonance cone in the type II region remain specific characteristics also of this particular HMM. Figure 5.12 shows the calculation of the resonance cone angle for both the two analyzed systems. In the insets, the supercollimation occurs for the two systems exactly where the resonance cone angle goes to zero. Moreover, both the systems show the typical resonance cone at 750 nm, falling for both of them in the type II region [32].

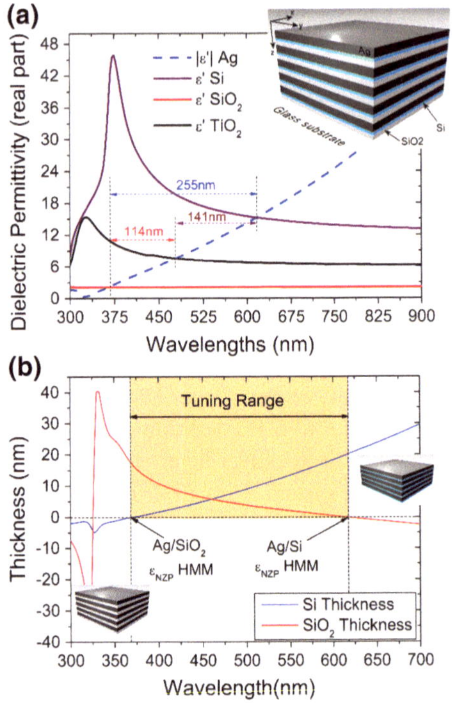

Fig. 5.9 **a** A comparison between the absolute value of the real part of the Ag dielectric permittivity (dashed blue curve) and that of Si, SiO_2 and TiO2. Three triplets are possible. In the inset, a realistic sketch of the SiO2/S*i*/Ag HMM considered here. **b** Si and SiO_2 thicknesses calculated via Eq. 5.22. At 365 nm the system degenerates into an Ag/SiO2 HMM, while at 620 nm the system falls into the Ag/Si bilayer HMM. Below 365 nm and above 620 nm the solutions are not physical. Reproduced with permission from Ref. [31] ©IOP Publishing

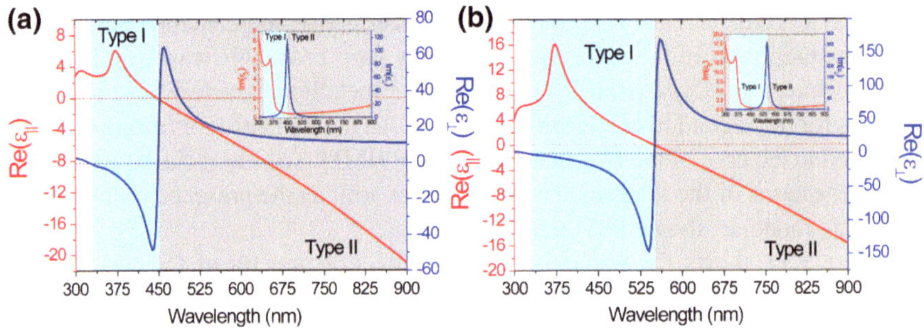

Fig. 5.10 Effective medium theory (EMT) of the two systems with the transition wavelength between the two coexisting anisotropies at 450 nm (**a**) and at 550 nm (**b**). The imaginary parts showing the typical Lorentzian shape that peaked exactly at the transition wavelengths are shown in the insets. Reproduced with permission from Ref. [31] ©IOP Publishing

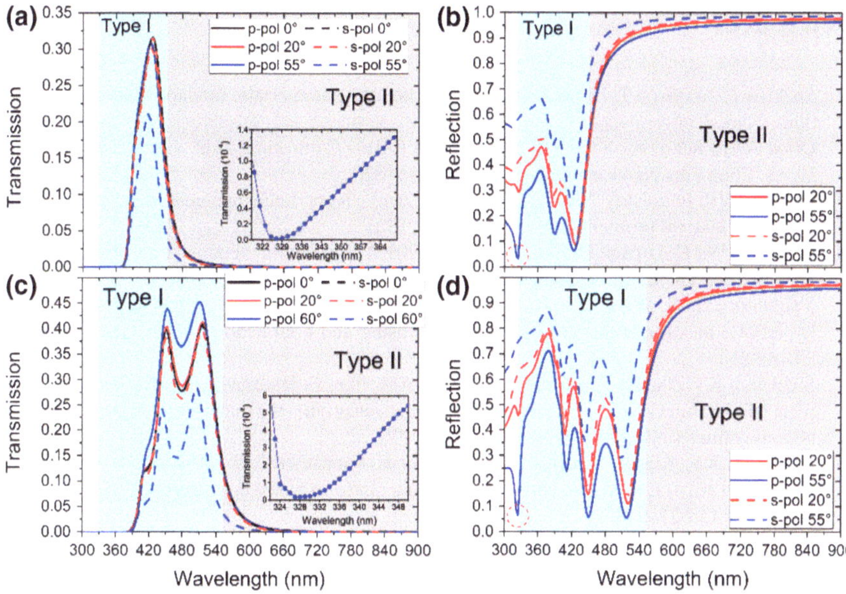

Fig. 5.11 Transmission and reflection spectra of the HMM with transition wavelength λtrans = 450 nm (**a**, **b**) and λtrans = 550 nm (**c**, **d**) showing a range of transparency within the *type I* region and high reflectivity within the type II range. The insets in **a** and **c** correspond to a zoom-in of the dips present at 327 nm, confirming the dielectric/type I transition, due to the presence of Ferrell–Berreman modes inside the structure. Reproduced with permission from Ref. [31] ©IOP Publishing

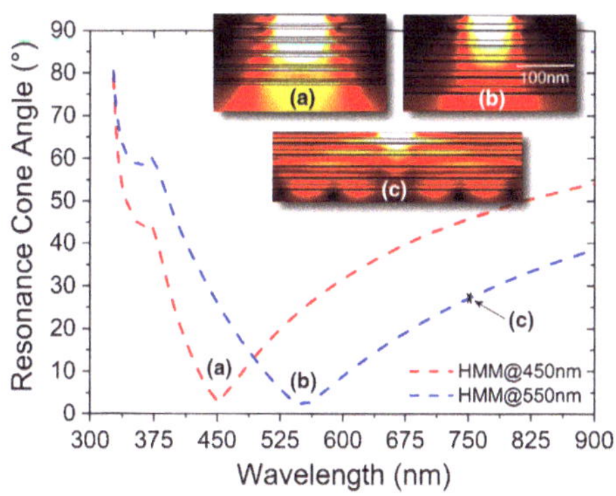

Fig. 5.12 Resonance cone angle calculation for both the HMM systems showing a minimum very close to zero in correspondence with the two transition wavelengths. In the insets are FEM simulations of the electric field propagation inside the structures showing a canalization effect at the respective canalization wavelengths, manifested as a straight subwavelength confined light beam. Adapted from [31]. ©IOP Publishing. Reproduced with permission. All rights reserved

References

1. Shekhar P, Atkinson J, Jacob Z (2014) Hyperbolic metamaterials: fundamentals and applications. Nano Converg 1:1–17
2. Orlov AA, Zhukovsky SV, Iorsh IV, Belov PA (2014) Controlling light with plasmonic multilayers. Photonics Nanostr Fundam Appl 12:213–220
3. Drachev VP, Podolskiy VA, Kildishev AV (2013) Hyperbolic metamaterials: new physics behind a classical problem. Opt Exp 21:1699–1701
4. Ferrari L, Wu C, Lepage D, Zhang X, Liu Z (2015) Hyperbolic metamaterials and their applications. Prog Quantum Electron 40:1–40
5. Vasilantonakis N, Wurtz GA, Podolskiy VA, Zayats AV (2015) Refractive index sensing with hyperbolic metamaterials: strategies for biosensing and nonlinearity enhancement. Opt Exp 23:14329
6. Sreekanth KV, Alapan Y, ElKabbash M, Wen AM, Ilker E, Hinczewski M, Gurkan UA, Steinmetz NF, Strangi G (2016) Enhancing the angular sensitivity of plasmonic sensors using hyperbolic metamaterials. Adv Opt Mater 4:1767–1772
7. Sreekanth KV, Alapan Y, ElKabbash M, Ilker E, Hinczewski M, Gurkan UA, De Luca A, Strangi G (2016) Extreme sensitivity biosensing platform based on hyperbolic metamaterials. Nat Mater 15:621–627
8. Caligiuri V, Dhama R, Sreekanth KV, Strangi G, De Luca A (2016) Dielectric singularity in hyperbolic metamaterials: the inversion point of coexisting anisotropies. Sci Rep 6:20002
9. Caligiuri V, Pezzi L, Veltri A, De Luca A (2017) Resonant gain singularities in 1D and 3D metal-dielectric multilayered nanostructures. ACS Nano 11:1012–1025
10. Simovski C, Maslovski S, Nefedov I, Kosulnikov S, Belov P, Tretyakov S (2015) Hyperlens makes thermal emission strongly super-Planckian. Photonics Nanostr Fundam Appl 13:31–41
11. Lee H, Liu Z, Xiong Y, Sun C, Zhang X (2007) Development of optical hyperlens for imaging below the diffraction limit. Opt Exp 15:1588
12. Bak AO, Giannini V, Maier SA, Phillips CC (2013) Super-resolution with a positive epsilon multi-quantum-well super-lens. Appl Phys Lett 103:261110
13. Jacob Z, Smolyaninov II, Narimanov EE (2012) Broadband Purcell effect: radiative decay engineering with metamaterials. Appl Phys Lett 100:181105
14. Jacob Z, Kim JY, Naik GV, Boltasseva A, Narimanov EE, Shalaev VM (2010) Engineering photonic density of states using metamaterials. Appl Phys B Lasers Opt 100:215–218
15. Noginov MA, Li H, Barnakov YA, Dryden D, Nataraj G, Zhu G, Bonner CE, Mayy M, Jacob Z, Narimanov EE (2010) Controlling spontaneous emission with metamaterials. Opt Lett 35:1863
16. Sreekanth KV, Krishna KH, De Luca A, Strangi G (2014) Large spontaneous emission rate enhancement in grating coupled hyperbolic metamaterials. Sci Rep 4:6340
17. Guo Y, Jacob Z (2013) Thermal hyperbolic metamaterials. Opt Exp 21:15014
18. Biehs SA, Tschikin M, Ben-Abdallah P (2012) Hyperbolic metamaterials as an analog of a blackbody in the near field. Phys Rev Lett 109:104301
19. Smolyaninov II (2013) Quantum electromagnetic "black Holes" in a strong magnetic field. J Phys G: Nucl Part Phys 40:015005
20. Smolyaninov II, Kildishev AV (2013) Light propagation through random hyperbolic media. Opt Lett 38:971
21. Kidwai O, Zhukovsky SV, Sipe JE (2012) Effective-medium approach to planar multilayer hyperbolic metamaterials: strengths and limitations. Phys Rev A 85:053842
22. Caligiuri V, Palei M, Imran M, Manna L, Krahne R (2018) Planar double-epsilon-near-zero cavities for spontaneous emission and Purcell effect enhancement. ACS Photonics 5:2287–2294
23. Dionne JA, Lezec HJ, Atwater HA (2006) Highly confined photon transport in subwavelength metallic slot waveguides. Nano Lett 6:1928–1932
24. Orlov AA, Voroshilov PM, Belov PA, Kivshar YS (2011) Engineered optical nonlocality in nanostructured metamaterials. Phys Rev B 84:045424
25. Castaldi G, Galdi V, Alù A, Engheta N (2012) Nonlocal transformation optics. Phys Rev Lett 108:063902

26. Naik GV, Saha B, Liu J, Saber SM, Stach EA, Irudayaraj JMK, Sands TD, Shalaev VM, Boltasseva A (2014) Epitaxial superlattices with titanium nitride as a plasmonic component for optical hyperbolic metamaterials. Proc Natl Acad Sci 111:7546–7551
27. Newman W, Cortes CL, Atkinson J, Pramanik S, DeCorby RG, Jacob Z (2014) Ferrell – Berreman modes in plasmonic epsilon-near-zero media. ACS Photonics 2:2–7
28. Ramakrishna SA, Pendry JB, Wiltshire MCK, Stewart WJ (2003) Imaging the near field. J Mod Opt 50:1419–1430
29. Fang N (2005) Sub-diffraction-limited optical imaging with a silver superlens. Science 308:534–537
30. Smalley JST, Vallini F, Zhang X, Fainman Y (2018) Dynamically tunable and active hyperbolic metamaterials. Adv Opt Photonics 10:354
31. Caligiuri V, De Luca A (2016) Metal-semiconductor-oxide extreme hyperbolic metamaterials for selectable canalization wavelength. J Phys D Appl Phys 49:08LT01
32. De Ceglia D, Vincenti MA, Campione S, Capolino F, Haus JW, Scalora M (2014) Phys Rev B - Condens Matter Mater Phys 89:1

Chapter 6
Resonant Gain Singularities in Hyperbolic Metamaterials

The interaction between dyes and plasmonic metals has been widely studied in numerous frameworks. Förster Resonant Energy Transfer (FRET) between molecular fluorescent donors and plasmonic acceptors has been widely demonstrated [1–3], and coherent resonant emission has been theorized and experimentally achieved in a regime called Surface Plasmon Amplification by Stimulated Emission of Radiation [4]. The Ohmic loss compensation via gain materials has been broadly described as well both in colloidal nano-particles and multilayered systems [5, 6]. In the previous chapter, a new propagation regime has been introduced in which a suitably dimensioned HMM can collimate light with unprecedented resolution. It has been explained that such a regime, called Epsilon-near-zero-and-pole (ENZP), is characterized by high losses in the visible. In this chapter, it will be shown how to compensate losses in the ENZP regime by means of a particular fluorescent blend designed appositely.

6.1 Resonant Gain Epsilon-Near-Zero and Pole Condition

As expressed in Chap. 5, HMMs can be modelled with a certain accuracy within the framework of the EMT. Even though we extensively explained the main reasons why a more sophisticated approach may be needed to describe HMMs, the occurrence of the so-called ENZP regime can be predicted by the simple approximation given by the EMT. One of the main advantages of this approach is that it is intrinsically Kramers-Krönig consistent, therefore interesting conclusions can be drawn also for the imaginary part of the effective dielectric permittivities of the HMM. An extended version of the typical equations of the EMA can give a thorough quick inspection on the features of the imaginary parts as well:

$$\tilde{\varepsilon}_{||} = \frac{\varepsilon_d^{'} t_d + \varepsilon_m^{'} t_m}{t_d + t_m} + i\frac{\varepsilon_d^{''} t_d + \varepsilon_m^{''} t_m}{t_d + t_m}; \tag{6.1}$$

Sreekanth K. V. et al., *New Directions in Thin Film Nanophotonics*, Progress in Optical Science and Photonics 6, https://doi.org/10.1007/978-981-13-8891-0_6

$$\tilde{\varepsilon}_{\perp} = \frac{\varepsilon'_m t_m\left[\left(\varepsilon'_d\right)^2 + \left(\varepsilon''_d\right)^2\right] + \varepsilon'_d t_d\left[\left(\varepsilon'_m\right)^2 + \left(\varepsilon''_m\right)^2\right]}{\left(\varepsilon'_d t_m + \varepsilon'_m t_d\right)^2 + \left(\varepsilon''_d t_m + \varepsilon''_m t_d\right)^2}(t_m + t_d)$$
$$+ i\frac{\varepsilon''_m t_m\left[\left(\varepsilon'_d\right)^2 + \left(\varepsilon''_d\right)^2\right] + \varepsilon''_d t_d\left[\left(\varepsilon'_m\right)^2 + \left(\varepsilon''_m\right)^2\right]}{\left(\varepsilon'_d t_m + \varepsilon'_m t_d\right)^2 + \left(\varepsilon''_d t_m + \varepsilon''_m t_d\right)^2}(t_m + t_d) \quad (6.2)$$

Here superscripts prime and double-prime refer to real and imaginary parts of $\tilde{\varepsilon}_m$ and $\tilde{\varepsilon}_d$, complex permittivities of metal and dielectric, respectively, while t_d and t_m are their thicknesses. As a first consideration, it can be noticed that even though the losses of the material do not play any role in the real part of $\tilde{\varepsilon}_{||}$ (for this component the real and imaginary parts are practically decoupled), they play a fundamental role in the real part of $\tilde{\varepsilon}_{\perp}$. Even in the case of very low loss dielectrics ($\tilde{\varepsilon}_d \approx 0$), the imaginary part of the metal's dielectric permittivity is able to heavily modify the value of $\tilde{\varepsilon}'_{\perp}$ and, therefore, the shape of the equifrequency contour. Equations (6.1) and (6.2) show that the imaginary part of $\tilde{\varepsilon}_{\perp}$ can become negative and show express a pole, just like in the case of the real parts. Understanding the role of a negative imaginary permittivity is not easy. However, since positive imaginary part of the dielectric permittivity is usually associated to an absorbent material, negative values of this quantity may be associated to a gain medium. Such a point of view has been fruitfully adopted in describing the dielectric permittivity of quantum emitters and dyes within their emission regime [7–9]. It is, therefore, plausible expecting that very high (ideally infinite) imaginary permittivity is related to very high gain values. In principle, if such a high-gain material is embedded into a cavity and enough feedback is provided, this system becomes a promising candidate for manifesting LASER emission. The condition for which the imaginary part of $\tilde{\varepsilon}_{\perp}$ shows a singularity, passing ideally from $+\infty$ to $-\infty$, goes under the name of resonant gain [10]. Here, it is interesting noticing that similar considerations have been drawn by Lawandy, in describing the possibility of reaching a singularity in the amplification of localized surface plasmons if a plasmonic nano-particle is immersed in a dye with suitable concentration [11].

As shown in Eqs. (6.1) and (6.2), $\tilde{\varepsilon}_{||}$ can manifest a zero exactly in correspondence of the pole of the imaginary part of $\tilde{\varepsilon}_{\perp}$ if these three conditions are satisfied simultaneously:

$$t_d = t_m; \quad (6.3)$$

$$\varepsilon'_m = \varepsilon'_d; \quad (6.4)$$

$$\varepsilon''_m = \varepsilon''_d; \quad (6.5)$$

It is worth noticing that the first two conditions naturally lead to the ENZP regime explained in Chap. 5. This is a very important point, since it frames the Resonant Gain (RG) regime as an exceptional case of ENZP HMM, in which losses are perfectly compensated. It is reasonable expecting that LASER emission eventually occurring in such a loss compensated HMM, must be intrinsically nano-collimated. It has already been explained in Chap. 5 how to satisfy conditions (6.3) and (6.4) for a single wavelength per metal/dielectric pair. The third condition expressed in Eq. (6.5) can instead be satisfied only by suitably engineering a gain material inside the HMM. For the critical value of $\varepsilon_d'' = \varepsilon_m''$, a complete compensation of losses is achieved and the RG-ENZP regime is reached.

A HMM fulfilling the conditions expressed in Eqs. (6.3)–(6.5) is called Resonant Gain ENZP HMM (RG-HMM), being the protagonist of this chapter. Straightforwardly, satisfying these three conditions, requires a particular dielectric blend in which a passive, high-index dielectric matrix is employed as a host for a highly efficient dye, providing the necessary gain.

We can define a roadmap for designing the suitable "gain blend" for achieving the dielectric singularity as follows:

(1) Selecting a high refractive index dielectric to be used as a host for the dye medium. This step fixes the ENZP wavelength, since the passive, high-index dielectric host matrix plays the major role in determining the real dielectric permittivity of the blend.
(2) Finding a gain material dye whose emission peak lies exactly at the ENZP wavelength, determined at Step 1.
(3) Calculating the value of ε_d'' for which $\varepsilon_\perp''$ shows a pole at the ENZP wavelength, determined at Step 1.
(4) Determining the concentration $\overline{N}_0$ of dye molecules to be embedded in the blend in order to obtain the value of ε_d'' calculated in Step 3.
(5) Calculating the refractive index of the blend.
(6) Verifying the effective optical response of the HMM made of a metal (for example Ag) and the dielectric blend designed in Step 3. This step verifies the occurrence of a singularity in $\varepsilon_\perp''$ at the ENZP wavelength found at Step 1.

6.2 Design of the Gain Blend

6.2.1 Step 1—*Selecting a High Refractive Index Dielectric*

It is clear that the engineering of a suitable fluorescent dielectric layer to be employed in the RG-HMM is of fundamental importance. As reported in Chap. 5, in order to push the ENZP condition in the visible range, a dielectric with very high real permittivity is required. TiO_2 constitutes a perfect candidate. At first, it holds the suitably high refractive index in the visible range and, on the other hand, it has been demonstrated to be a versatile platform for the controlled embedding of many dopants. There are indeed several approaches by which TiO_2 can be functionalized in order to embed fluorescent molecules. One consists in a slightly modified Sol-Gel synthesis of TiO_2, involving also a polymeric matrix [12]. Very promising approaches are also the so called *grafting-to* and *grafting-from* techniques. In the former, a dye-doped polymer chain reacts with a TiO_2 nanoparticle, while in the latter, a polymer chain is directly grown on the top of the TiO_2 nanoparticle and successively functionalized with the dye [13, 14].

The Ag/TiO_2 HMM considered in this chapter manifests an ENZP condition at 426 nm. This feature is readily verifiable following the procedure shown in Chap. 5.

6.2.2 Step 2—*Selecting a Dye with Emission Peaked at 426 nm*

No particular constraints are required for the nature of the fluorophore, provided the suitable efficiency and correct emission are considered. This opens, for example, to the realm of novel inorganic fluorophores like perovskites and quantum dots [15–17]. These materials demonstrated very narrow photoluminescence, excellent quantum yield, very low ASE amplified spontaneous emission (ASE) threshold and completely tailorable emission wavelength. This last feature lifts from the lack of choice introduced by typical organic fluorophores. However, one of the drawbacks of inorganic dyes are their typical radiative decay lifetimes that, reaching the tens of nanoseconds in most of the cases, are too long for such applications. Therefore, in this chapter we will consider Coumarine C500 as the fluorescent material to be embedded in the blend. As a high dielectric host, TiO_2 will be considered. It should be also considered that, once the dye is placed in proximity or even inside a plasmonic material, its decay lifetime changes dramatically, since the density of photonic states results highly modified by the additional radiative relaxation channels introduced by the plasmonic system. In a HMM such a phenomenon is even more evident, since the so-called *high-k* modes described in Chap. 5 constitute very favorable relaxation channels, providing noticeable enhancement of the decay rate and very high Purcell effect [18–20].

6.2.3 Step 3—*Calculating the Value of ε''_d for Which $\varepsilon''_\perp$ Shows a Pole at 426 nm*

Once the ENZP wavelength is fixed, the value of ε''_d is intrinsically determined, independently on the selected dielectric, by Eq. (6.5). In this case, as shown in Fig. 6.1, $\varepsilon''_d = -0.54$. At this concentration, $\varepsilon''_\perp$ shows a singularity and its value becomes negatively very high. Counterintuitively, higher ε''_d lead to lower $\varepsilon''_\perp$. Once more, it is important highlighting that such a condition is a property of the HMM and does not depend on the dielectric.

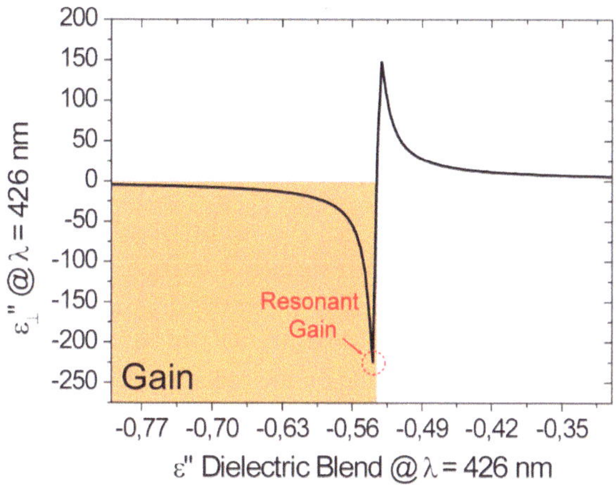

Fig. 6.1 Imaginary effective perpendicular dielectric permittivity calculated in the framework of the EMT at 426 nm, being the ENZP wavelength of a HMM made of Ag and of a dielectric blend embedding TiO_2, as a function of the imaginary permittivity of the dielectric blend. The resonant value of the imaginary permittivity of the dielectric blend is highlighted in red. Reproduced with permission from Ref. [10]. Copyright (2019) American Chemical Society

6.2.4 Step 4 *and* 5—*Calculation of the Concentration* $\overline{N}_0$ *of Dye Molecules and of the "Gain Blend" Effective Permittivity*

Once the fundamental passive and active components are selected, the dielectric permittivity $\tilde{\varepsilon}_d = \varepsilon_d^{'} + i\varepsilon_d^{''}$ of the blend can be designed as reported by Campione et al. [7]:

$$\tilde{\varepsilon}_g = \varepsilon_0\varepsilon_r + \frac{\sigma_a}{\omega^2 + i\Delta\omega_a\omega - \omega_a^2}\frac{(\tau_{21} - \tau_{10})\Gamma_{pump}}{(\tau_{32} + \tau_{21} + \tau_{10})\Gamma_{pump}}\overline{N}_0 \tag{6.6}$$

where ω is the angular frequency, ω_a is the emitting angular frequency, $\Delta\omega_a$ is the linewidth of the dye transition, ε_0 is the vacuum dielectric permittivity, $\varepsilon_r(\omega)$ is the frequency dependent complex permittivity of the host dielectric and σ_a is a coupling strength parameter [21]. $\overline{N}_0$ represents the total dye concentration expressed as $\overline{N}_0 = N_0(r) + N_1(r) + N_2(r) + N_3(r)$ being N_i the occupation density of the ith state, whereas Γ_{pump} the pump rate and $\tau_{i+1,i}$ the relaxation time during the $(i+1)$th to ith state transition. In our case, the following values have been considered:

- $\lambda = 2\pi c/\omega$ spanning from 300 to 900 nm.
- $\lambda_a = \frac{2\pi c}{\omega_a} = 426$ nm, being the ENZP transition wavelength for the considered HMM. This wavelength corresponds also to the emission peak of Coumarine C500.
- $\Delta\omega_a = 2\pi\Delta\upsilon_a = \frac{2\pi\Delta\lambda_a}{\lambda_a^2} = 207.7$ THz (with $\Delta\lambda_a = 20$ nm).
- $\sigma_a = 6\pi\varepsilon_0 c^3\frac{\eta}{(\tau_{21}\omega_a^2\sqrt{\varepsilon_r})}$ (C^2/kg).

- $\eta = 0.78$.
- $\tau_{10} = \tau_{32} = 100$ fs.
- $\tau_{21} = 110$ ps.
- $\Gamma_{pump} = \frac{\sigma_{abs} I}{hf}$ (s^{-1}), where h is the Planck constant.

The intensity I is expressed as $I = \frac{P}{A} = \frac{E}{\tau_{pulse} A}$, in which the pump energy $E = 10$ μJ and the pulse repetition rate $f = 80$ MHz are those of a mode-locked Ti:Sapphire pulsed laser. The pumped surface A is equal to πr^2 with $r = 2$ mm, typical of an unfocused laser beam. The pump rate Γ_{pump} directly depends on the absorption cross section of the proposed dyes, that can be recovered in Ref. [21] and results to be $\sigma_{abs} = \frac{\gamma_{rad} \lambda_a^2}{2\pi \Delta\omega_a}$, depending on the radiative decay rate $\gamma_{rad} = \frac{\eta}{\tau_{21}}$ from level 2 to level 1.

The total number of dye molecules $\overline{N}_0$ at the ground state determines the value $\varepsilon_d^{''}$ of the imaginary part of the dielectric permittivity of the blend. It is important to recall that here we want to dimension $\overline{N}_0$ in order to obtain $\varepsilon_d^{''} = -0.54$ that, as determined in the previous step, is the value at which $\varepsilon_\perp^{''}$ shows a singularity. By fixing the wavelength at 426 nm, the imaginary permittivity of the dielectric blend can be swept by changing $\overline{N}_0$ and the obtained values can be readily put into Eq. (6.2), to find the concentration $\overline{N}_0$ that induces the singularity in $\varepsilon_\perp^{''}$. As shown in Fig. 6.2a, the RG regime occurs for $\overline{N}_0 = 6.55 \times 10^{18}$ cm^{-3}. Once $\overline{N}_0$ is fixed, it is possible to calculate the effective dielectric permittivity of the "*gain blend*", via Eq. (6.5). Figure 6.2b, c illustrate, respectively, the real and imaginary part of the gain blend as calculated via Eq. (6.5) for the resonant dye molecules concentration and in the case of a doubled concentration. This last condition will allow demonstrating that, as mentioned before, a higher negative $\varepsilon_d^{''}$ (higher $\overline{N}_0$) leads to a lower $\varepsilon_\perp^{''}$ value.

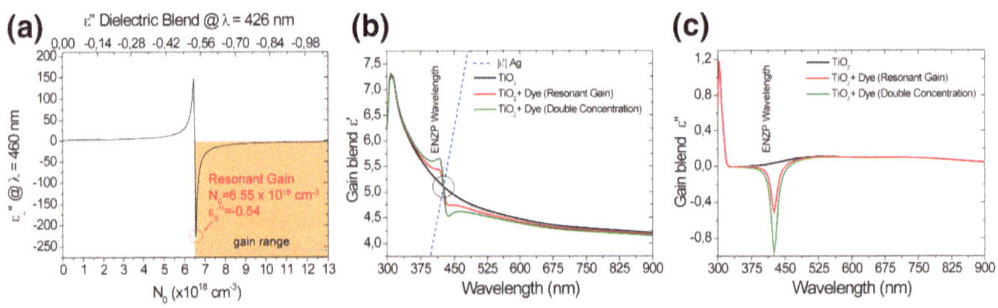

Fig. 6.2 **a** Imaginary effective perpendicular dielectric permittivity at 426 nm calculated as a function of the N_0 dye molecules concentration (imaginary dielectric permittivity) embedded in the dielectric blend. A resonant concentration equal to 6.55×10^{18} cm^{-3} is found, while increasing N_0 leads to a lowering of $\varepsilon_\perp^{''}$. **b** Real and **c** imaginary dielectric permittivity of the dielectric blend designed by means of Eq. (6.5), in the cases of no dye (black), resonant gain concentration (red) and double concentration (green), compared to the absolute value of the real dielectric permittivity of Ag (blue dashed curves). Noticeably, the crossing point between Ag and all the curves occurs always at the ENZP wavelength, coinciding with the Lorentzian peak of the dielectric blend. Reproduced with permission from Ref. [10]. Copyright (2019) American Chemical Society

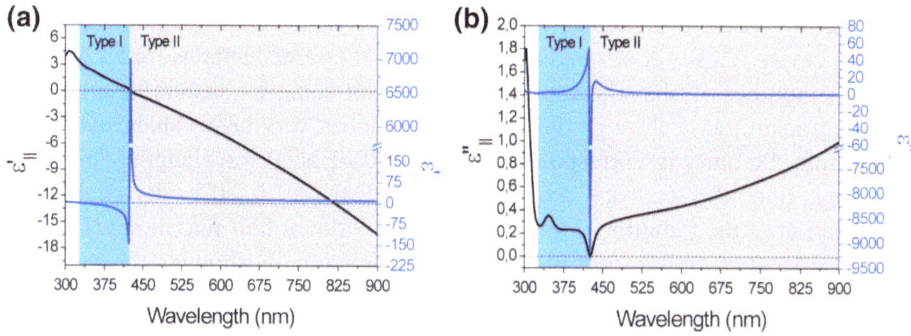

Fig. 6.3 **a** Real and **b** imaginary effective dielectric permittivity calculated by means of the EMT for a RG-HMM. A resonance is clearly visible in both the real and imaginary part of $\varepsilon_{\perp}$, being the signature of the RG-ENZP regime. Reproduced with permission from Ref. [10]. Copyright (2019) American Chemical Society

6.2.5 Step 6—*Verifying the Presence of the "Resonant Gain Singularity" in $\varepsilon''_{\perp}$ at the ENZP Wavelength*

Once the gain blend is dimensioned, the complete RG-HMM can be designed in the framework of the EMT. In particular, we are looking for a singularity in both the imaginary and the real part of the effective perpendicular dielectric permittivity. Figure 6.3a, b show the real and imaginary part of the parallel and perpendicular dielectric permittivity calculated by means of the effective medium theory for a 50% fill fraction HMM embedding a dielectric designed by means of Eq. (6.5), considering $\overline{N}_0 = 6.55 \times 10^{18}$ cm^{-3}. A singularity is present at 426 nm, being the predicted ENZP, in both the real and imaginary part of $\varepsilon_{\perp}$.

6.3 Supercollimation and Light Amplification in the RG-HMM

In Chap. 5, it has been explained how the supercollimation does not depend on the number of layers by which the HMM is made of but only on the fill fraction which has to be 50%. However, when looking for light amplification and LASER effect, it is important ensuring a feedback path to the involved gain material, in order to overcome the losses and reach the LASER threshold. For this reason, in the RG-HMM configuration, the number of layers is very important.

Figure 6.4a shows the transmittance calculated via SMM method of a RG-HMM made of 20 nm Ag layers, alternated to 20 nm of dielectric blend, dimensioned as shown before. Interestingly, the number of bilayers that optimizes the transmittance can be clearly identified in 12. With 12 bilayers the distance between the first and the last Ag layer results about 420 nm, then comparable to the RG-ENZP wavelength of the HMM. Therefore, we can conclude that the HMM itself plays the role of a resonant cavity. This is confirmed by the fact that the same happens for 24 bilayers. When the number of bilayers reaches the minimum value for sustaining amplification, the transmittance through the RG-HMM reaches very high values, eventually overcoming the unity in correspondence of the RG-ENZP wavelength. This is not surprising, since the feedback provided by the HMM as a self-cavity enables the amplification of the radiation. Moreover, the full width at half-maximum (FWHM) of the transmission peak results highly narrowed, eventually showing the characteristics of monochromatic LASER emission. Figure 6.4b shows this scenario by means of SMM simulations. The HMMs made of 12 and 24 bilayers enable overcoming the unitary transmittance at the RG-ENZP wavelength, while the one calculated for a 6 bilayers RG-HMM (black curves), as well as that of a 12 bilayers HMM in which the dye concentration in the dielectric is doubled (green curve) is always much lower than 1. This last phenomenon allows demonstrating what expressed in Figs. 6.1 and 6.2a: overcoming the RG dye concentration leads to a lowering of the emission, counterintuitively. More insight on the reason why this happens can be found after a deep investigation of the EMT homogenized parameters of the dye doped HMM.

Figure 6.4c shows that a dramatic topological transition can be induced by modifying the concentration of dye molecules in the dielectric. In particular, the HMM manifests a *type II* anisotropy when the concentration is lower than the singular one, holding growing positive perpendicular dielectric permittivity the closer to the resonant concentration. This parameter undergoes an abrupt inversion right after the resonant concentration, switching the effective response to that of an effective metal. Such a transition is crucial, since it has been explained in Chap. 5 that the effective metal does not allow any propagating solution in the space of the real wavevectors, meaning that no radiative propagation is allowed in this regime. This leads to the consequence that the metamaterial acting in this regime allows no propagation through it, neither in the form of resonant cones, resulting completely reflective. For this reason, no propagation (and, of course amplification) is allowed at the ENZP wavelength, regardless of the dye concentration in the dielectric layer.

Figure 6.5 reports the analysis of the light propagation in (a, d) undoped, (b, e) RG concentration and (c, f) with a doubled concentration with respect to the resonant one. As can be seen, the canalization effect is amplified by the presence of the resonant amount of gain, demonstrating that the frequency corresponding to the ENZP propagates through the structure as a well collimated solitary wave (soliton), with a diameter comparable to the period of the HMM, being the thickness of its metal/dielectric unit cell. Indeed, as shown in the x-cuts of the electric field taken at the exit layer of the HMM, for $x = 40$ nm and the ENZP wavelength being 426 nm, a more than $\lambda/10$ collimation is reached. Moreover, due to the RG condition, the increase of light intensity confined inside the structure is clearly visible as a well collimated and amplified soliton. It is important noticing that the RG condition improves the canalization performances of the ENZP HMM, since no broadening of the soliton due to losses mismatch is introduced. In the case of a HMM embedding a doubled amount of gain with respect to the resonant one (Fig. 6.5c, f), the light intensity propagating inside the HMM is extremely attenuated (see the cuts at the output of HMM), confirming the behavior shown in Fig. 6.5a, b.

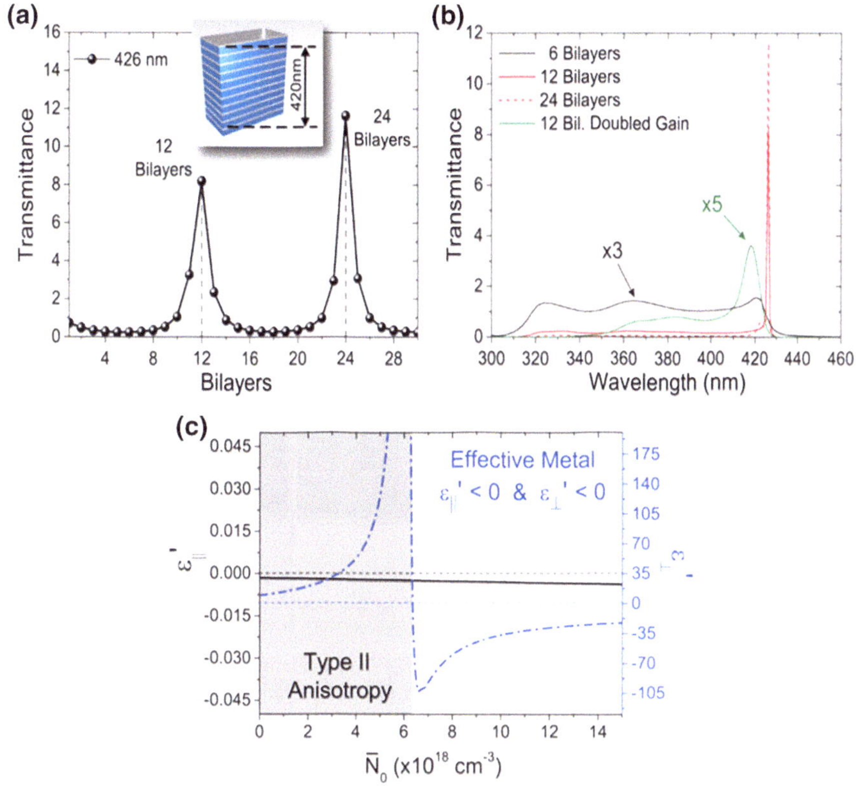

Fig. 6.4 **a** Transmittance calculated via SMM at 426 nm for the RG-HMM, as a function of the number of bilayers by which the HMM is made of. In correspondence of multiples of 12 bilayers, the whole HMM behaves as a cavity for the light emitted by the dye and provides the necessary feedback to trigger the amplification process. **b** Transmittance spectra calculated via SMM at 12 (solid red) and 24 (dashed red) bilayers in comparison with that of a 6 bilayer RG-HMM (black) and 12 bilayers doubled-concentration HMM (green). Transmittance overcomes unity only in the case of RG-HMMs made of multiples of 12 bilayers, reaching very narrow emission. **c** Real parallel (solid black) and perpendicular (dot-dashed blue) effective dielectric permittivity calculated via EMT at 426 nm as a function of the N_0 dye molecules concentration. A topological transition is induced when increasing the amount of gain in the dielectric blend. When N_0 is higher than the resonant value, an effective metal anisotropy is induced. In this regime, no real wavevectors and, consequently, no propagating waves are allowed neither in the form of resonance cones. As a result, over doping the HMM prevents light propagation through it. Reproduced with permission from Ref. [10]. Copyright (2019) American Chemical Society

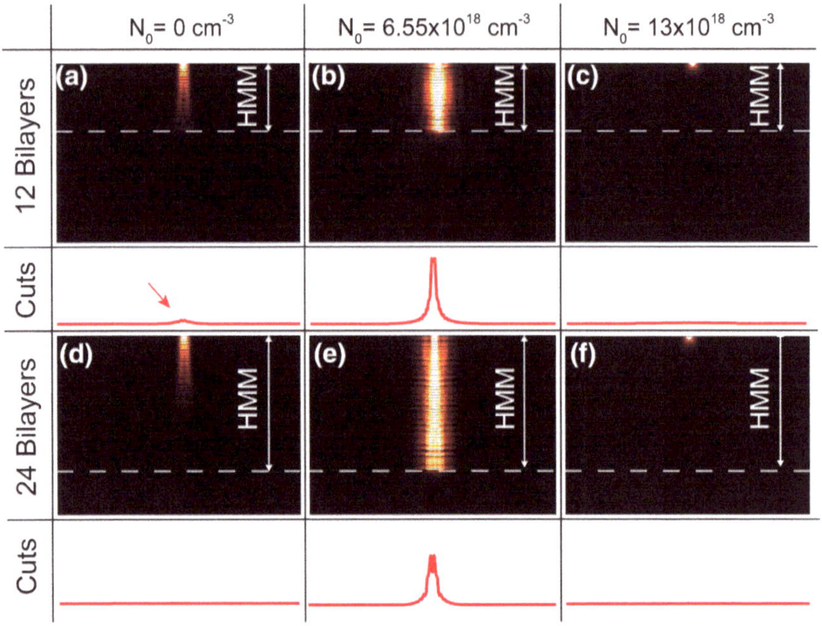

Fig. 6.5 FEM based simulations of the RG-ENZP HMM made of **a**–**c** 12 bilayers and **d**–**f** 24 bilayers, together with x-cuts showing the norm of the electric field at the exit of the HMMs. A noticeable enhancement of light propagation within the HMM as well as a remarkable collimation is obtained in the case of the RG-HMM (**b**, **e**), in comparison to the strongly attenuated passive ENZP HMM (**a**, **d**) and to the propagation forbidden-double-concentration HMM (**c**, **f**). Reproduced with permission from Ref. [10]. Copyright (2019) American Chemical Society

6.4 Self-Amplified Perfect Lens (APL)

One of the main applications of an ENZP HMM is the so-called perfect lens. Its working principle has been theoretically explained and experimentally demonstrated in Chap. 5. It has been outlined how such a configuration allows a resolution limited only by the period of the designed HMM, making it an ideal candidate for application as single molecule imaging, deeply subwavelength laser-beam lithography, ultrafine biosensing, and so on. Anyway, two weaknesses have to be improved. The first one is inherent to the design of the ENZP HMM. Indeed, as it can be seen in Eq. (6.2), the sharpness of the resonance is determined by the amplitude of the denominator that has to be as close as possible to zero. As shown previously, the RG-ENZP HMM configuration intrinsically solves the problem, providing an almost zero denominator optimizing the collimation performances of the perfect lens. The second issue is inherent to the losses. A perfect lens based on a classic ENZP HMM introduces losses by definition imposing a limitation to the amount of light intensity detectable at the exit of the device. The amplification properties of the RG-ENZP HMM reveal once more useful by automatically fixing this limitation.

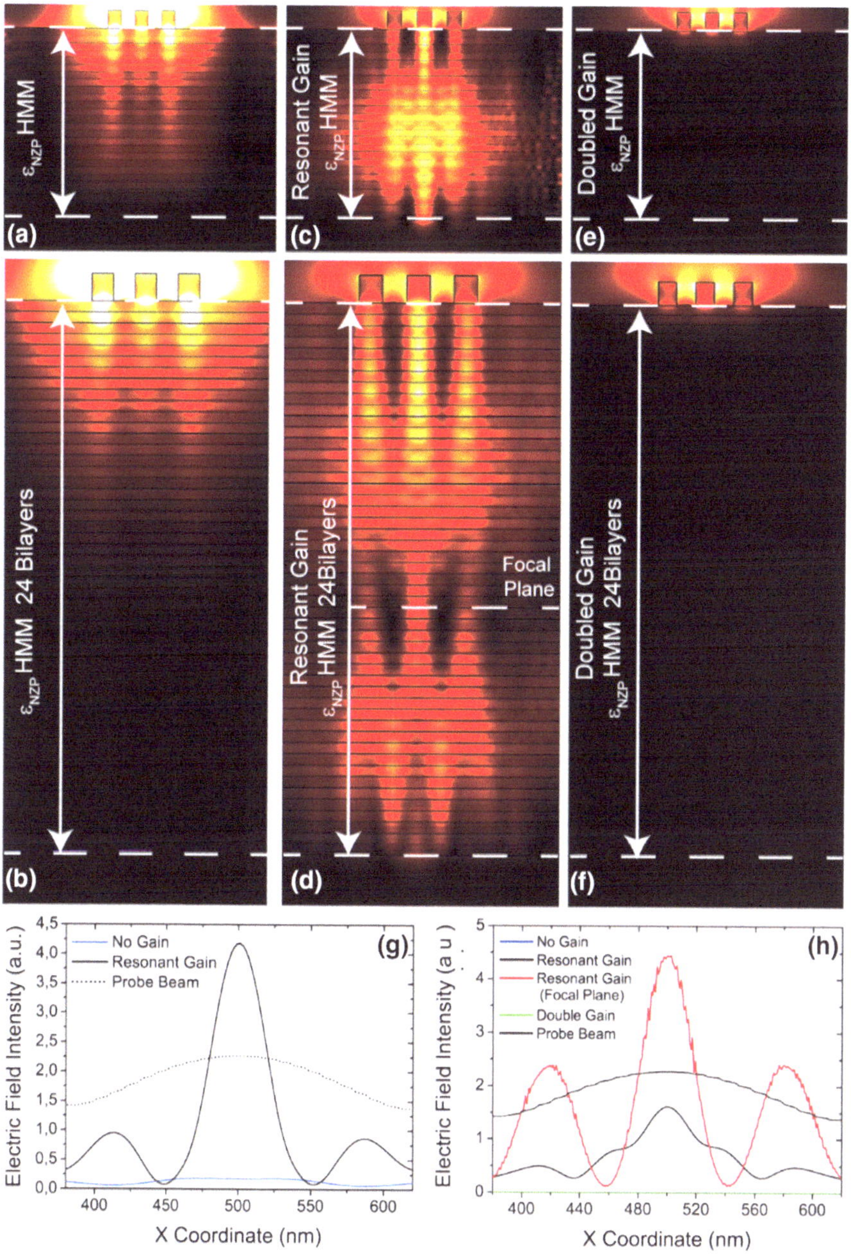

(a)
εNZP HMM
(c)
Resonant Gain
εNZP HMM
(e)
Doubled Gain
εNZP HMM
(b)
εNZP HMM 24 Bilayers
(d)
Resonant Gain
εNZP HMM 24Bilayers
Focal
Plane
(f)
Doubled Gain
εNZP HMM 24Bilayers
(g)
No Gain
Resonant Gain
Probe Beam
Electric Field Intensity (a.u.)
X Coordinate (nm)
(h)
No Gain
Resonant Gain
Resonant Gain
(Focal Plane)
Double Gain
Probe Beam
Electric Field Intensity (a.u.)
X Coordinate (nm)

◀**Fig. 6.6** FEM demonstration of the perfect imaging of three deeply subwavelength polymeric elements, placed on the top of 12-bilayers (**a**, **c**, **e**) and 24-bilayers (**b**, **d**, **f**) HMMs used as perfect lenses. Three cases are considered. (i) Passive ENZP HMM (**a**, **b**): in this case the supercollimation is present but the attenuation is very high and no nanometric objects are visible at the exit plane of the HMM. (ii) Resonant gain concentration (**c**, **d**): in this case in both the 12 bilayers and 24 bilayers RG-HMMs allow the individuation of the three elements, but only the 24 bilayers enables a noticeable amplification of the radiation, over the intensity of the probe beam. This is far more appreciable in the x-cuts of where the norm of the electric field at the exit of the RG-HMM is collected for both the **g** 12 bilayers and **h** 24 bilayers RG-HMM. (iii) Double concentration HMM (**e**, **f**): in this case no propagation inside the HMM is allowed, despite the high dye concentration. Reproduced with permission from Ref. [10]. Copyright (2019) American Chemical Society

Figure 6.6 shows two simulated amplified perfect lenses (APLs), based on 12- and the 24-bilayer RG-ENZP HMMs. Three deeply subwavelength polymer (PMMA) elements are placed on the top of the structures. Illuminating at the ENZP wavelength, the three elements are clearly detectable at the exit of the RG-ENZP HMM. A comparison between: (i) a classic, nondoped, perfect lens (Fig. 6.6a for the 12-bilayer HMM and Fig. 6.6b for the 24-bilayer HMM), (ii) the RG-ENZP (Fig. 6.6c for the 12 bilayer HMM and Fig. 6.6d for the 24bilayer HMM), and (iii) over doped ENZP HMM (Fig. 6.6e for the 12 bilayer HMM and Fig. 6.6f for the 24 bilayer HMM) is reported. An increased output light intensity is obtained in the case of both the 12- and 24-bilayer APLs (Fig. 6.6c, d) with respect to both the classic ENZP HMM and the over doped one. As expressed beforehand, the double-doped HMM undergoes a topological transition, switching from a *type II* to an effective metal HMM. Such a transition leads the HMM to be extremely reflective and, consequently, the field inside the structure results extremely attenuated. The x-coordinate cuts took at the exit layer of the HMM for the 12 bilayers HMM (Fig. 6.6g) and 24 bilayers HMM (Fig. 6.6h) confirm this phenomenon. Moreover, it has been demonstrated in Ref. [22] that a classic HMM can focus a parallel light rays bundle inside the structure. Such a property can be found also in the proposed 24 bilayer APL. Exactly in the middle a focal plane can be identified, in which not only are the three nanometric elements posed on the top of the RG-ENZP HMM perfectly resolved, but all of them are amplified with respect to the pump beam (see Fig. 6.6g).

References

1. Kochuveedu ST, Son T, Lee Y, Lee M, Kim D, Kim DH (2014) Revolutionizing the FRET-based light emission in core-shell nanostructures via comprehensive activity of surface plasmons. Sci Rep 4:4735
2. ElKabbash M, Rashed AR, Sreekanth KV, De Luca A, Infusino M, Strangi G (2016) Plasmon-exciton resonant energy transfer: across scales hybrid systems. J Nanomater 20:1–21
3. De Luca A, Dhama R, Rashed AR, Coutant C, Ravaine S, Barois P, Infusino M, Strangi G (2014) Double strong exciton-plasmon coupling in gold nanoshells infiltrated with fluorophores. Appl Phys Lett 104:103103

4. Noginov MA, Zhu G, Belgrave AM, Bakker R, Shalaev VM, Narimanov EE, Stout S, Herz E, Suteewong T, Wiesner U (2009) Demonstration of a spaser-based nanolaser. Nature 460:1110–1112
5. Noginov MA, Podolskiy VA, Zhu G, Mayy M, Bahoura M, Adegoke JA, Ritzo BA, Reynolds K (2008) Compensation of loss in propagating surface plasmon polariton by gain in adjacent dielectric medium. Opt Exp 16:1385
6. Krishna KH, Sreekanth KV, Strangi G (2016) Dye-embedded and nanopatterned hyperbolic metamaterials for spontaneous emission rate enhancement. J Opt Soc Am B 33:1038
7. Campione S, Albani M, Capolino F (2011) Complex modes and near-zero permittivity in 3D arrays of plasmonic nanoshells: loss compensation using gain. Opt Mater Exp 1:1077
8. Strangi G, De Luca A, Ravaine S, Ferrie M, Bartolino R (2011) Gain induced optical transparency in metamaterials. Appl Phys Lett 98:251912
9. De Leon I, Berini P (2010) Amplification of long-range surface plasmons by a dipolar gain medium. Nat Photon 4:382–387
10. Caligiuri V, Pezzi L, Veltri A, De Luca A (2017) Resonant gain singularities in 1D and 3D metal-dielectric multilayered nanostructures. ACS Nano 11:1012–1025
11. Lawandy NM (2004) Localized surface plasmon singularities in amplifying media. Appl Phys Lett 85:5040–5042
12. Caligiuri V, Lento R, Ricciardi L, Termine R, La Deda M, Siprova S, Golemme A, De Luca A (2018) Environmental control of the topological transition in metal/photoemissive-blend metamaterials. Adv Opt Mater 6:1–8
13. Gherardi F, Colombo A, D'Arienzo M, Di Credico B, Goidanich S, Morazzoni F, Simonutti R, Toniolo L (2016) Efficient self-cleaning treatments for built heritage based on highly photoactive and well-dispersible TiO_2 nanocrystals. Microchem J 126:54–62
14. Kumar SK, Jouault N, Benicewicz B, Neely T (2013) Nanocomposites with polymer grafted nanoparticles. Macromolecules 46:3199–3214
15. Imran M, Caligiuri V, Wang M, Goldoni L, Prato M, Krahne R, De Trizio L, Manna L (2018) Benzoyl halides as alternative precursors for the colloidal synthesis of lead based halide perovskite nanocrystals. J Am Chem Soc 140:2656–2664
16. Alivisatos AP (1996) Semiconductor clusters, nanocrystals, and quantum dots. Science 271:933–937
17. Grim JQ, Christodoulou S, Di Stasio F, Krahne R, Cingolani R, Manna L, Moreels I (2014) Continuous-wave biexciton lasing at room temperature using solution-processed quantum wells. Nat Nanotechnol 9:891–895
18. Noginov MA, Li H, Barnakov YA, Dryden D, Nataraj G, Zhu G, Bonner CE, Mayy M, Jacob Z, Narimanov EE (2010) Controlling spontaneous emission with metamaterials. Opt Lett 35:1863
19. Majumdar A, Xu X (2015) Monolayer semiconductor nanocavity lasers with ultralow thresholds. Nature 520:69–72
20. Sreekanth KV, Krishna KH, De Luca A, Strangi G (2014) Large spontaneous emission rate enhancement in grating coupled hyperbolic metamaterials. Sci Rep 4:6340
21. Siegman EA (1986) Lasers, Mill Valley, Calif. University Science Books
22. Lu D, Liu Z (2012) Hyperlenses and metalenses for far-field super-resolution imaging. Nat Commun 3:1205

Chapter 7
Metal/Photoemissive-Blend Hyperbolic Metamaterials for Controlling the Topological Transition

In this chapter, we present a simple method to overcome the intrinsic lack of tunability and reconfigurability manifested by HMMs. The proposed method bases on the engineering of a hygroscopic dielectric material consisting in a TiO_2/Polymer/Dye blend, providing a thermally tunable response in the visible range. Each of the blend components plays a fundamental role in the design of the optical, environmental and photophysical response of the complete HMM. In the following, it will be described the design of the thermo-responsive blend, the optical and photophysical characterization of the complete HMM embedding it as a dielectric as well as the thermal tuning of its optical properties. The presence of the dye within the dielectric matrix allows a visual detection of the thermal reconfiguration, permitting the proposed architecture to be considered as the proof-of-concept for the future engineering of a temperature sensor device working in a humidity-controlled environment, or vice versa, as a humidity sensor at a fixed temperature, both based on the proposed phenomenon.

7.1 Introduction

In Chap. 5, it has been discussed how the dynamic tuning of the optical response of the HMM still represents a major challenge preventing this technology to express its full potential. The reason is that the most common electrically tunable dielectrics like liquid crystals tend to lose their characteristics when squeezed at the nanoscale. For example, aligning thin layers of few tenths of nanometers of liquid crystals is challenging and the resulting anchoring forces of the mesogens would be too high to allow a reorientation of the molecular director with reasonably low voltages. Moreover, the alignment of the LCs is usually achieved by means of nanometrical layers of suitably deposited oxides that, when embedded in HMMs, contribute dramatically to their optical response eventually moving far from the $t_d = t_m$ condition [1, 2]. Electrical tuning is, however, still possible via field-effect [3]. Such a technique exploits the properties of a metal/oxide Schottky-like junction in which, under a suitable polarization, the depletion region extends in the oxide modifying its refrac-

Sreekanth K. V. et al., *New Directions in Thin Film Nanophotonics*, Progress in Optical Science and Photonics 6, https://doi.org/10.1007/978-981-13-8891-0_7

tive index. Even though such a regime has been theorized, it has not been verified experimentally. Always in Chap. 5 a new configuration based on three materials fundamental unit has been outlined in order to overcome the lack of design flexibility of ENZP HMMs [4]. Even though this configuration allows a fine design of the ENZP wavelength in a broad range, it still does not allow the dynamic tuning of the HMM, so that once the structure is fabricated there is no chance of modulating its optical response. The achievement of dynamic tunability of the optical properties of HMMs represents, therefore, a major challenge to face.

7.2 Design, Fabrication and Characterization of the Thermo-Responsive Blend

The expressions derived in Chap. 5 for the parallel and perpendicular effective permittivity of the HMM lead to the conclusion that a dynamic tunability of the dielectric permittivity of either the dielectric layer or the metal leads to a tuning of the optical response of the overall metamaterial. Even though at a first glance it might look straightforward accomplishing to this task by tapping into the plethora of tunable dielectrics, the nanometric thickness required for forming a HMM forces at excluding the most common of them like liquid crystals. An "ad hoc" approach is then needed. In this chapter, it will be explained the case of a suitable engineered dielectric blend whose refractive index is tailored to be very sensitive to changes in environmental factors. A novel dielectric material, consisting of a blend of TiO_2/Polymer/Dye (TPD), is manufactured in such a way that its optical response is sensitive to temperature and humidity. In the TPD blend, TiO_2 was chosen as the basic component due to its high refractive index and its transparency in the visible range. Thanks to this, when the TPD blend is embedded in a HMM, the ENZ wavelength occurring in the effective parallel dielectric permittivity falls in the visible range. Moreover, TiO_2 offers the possibility of a versatile sol-gel synthesis whose recipe can be modified in order to embed different dopants, crucial for reaching the desired properties. Thin TiO_2 films were therefore prepared by a sol-gel method, starting from titanium tetraisopropoxide (Ti $(OC_3H_7)_4$) as a precursor. The classic recipe was slightly modified introducing polyvinylpyrrolidone (PVP) and an organic dye (Coumarin C522B), both soluble in ethanol.

The use of PVP ensures a fine control of the refractive index of the blend. By acting on its concentration in the final blend, it is indeed possible to tune the topological transition (i.e. the ENZ wavelength) of the HMM with the photoluminescence peak of C522B, chosen because of its compatibility with the sol-gel protocol and remarkable quantum yield.

For PVP—(and all the other dye-doped solutions described in the chapter), 2% w/w of PVP 10,000 and/or 3% w/w of Coumarin 522B (with respect to Ti $(OC_3H_7)_4$) was added to the HCl/ethanol solution. The final TPD solution was spread on a clean glass substrate and spin coated at 3000 RPM for 1 min.

From the combination of the three elements by which the final blend is composed, many different dielectric blends can be achieved, whose real and imaginary refractive indices obtained via Spectroscopic Ellipsometry, are plotted in Fig. 7.1a, b:

1. Unsintered TiO_2+ PVP (TP): This combination manifests a refractive index slightly lower than that of the pure sintered TiO_2 (black dash-dotted lines). This is somehow expected, since the inclusion of the low index PVP reduces the effective refractive index. This blend already holds the tunability properties of the final TPD compound but does not allow a visual detection of the phenomenon and does not provide the possibility of a fine detection of the topological transition switch sensed via the weak coupling of the final HMM with the fluorescent molecules.
2. Unsintered TiO_2+ Dye (TD): This blend allows to explore the interaction between TiO_2 and the selected dye (C522b). The obtained films do not show a good morphology and, as a general consideration, the close interaction between TiO_2 and C522b results detrimental for the photophysical performances of the dye. Nevertheless, such a combination is crucial to attribute the thermo-chromic effect to the interaction between TiO_2 and dye, as will be shown after in the chapter. It is important noticing that the presence of the very low index C522b dye significantly lowers the refractive index of the overall blend, as shown in Fig. 7.1a (green, dotted curve).
3. Complete TiO_2+ PVP + Dye (TPD): This combination holds all the thermo-chromic characteristics of the TD compound but, due to the presence of the dye, its refractive index is close to that of the TP blend. This characteristic result crucial in moving the topological transition of the HMM employing it as a dielectric material within the photoluminescence range of the dye.

The Lorentzian peak occurring in the imaginary part k of the TD's and TPD's blends (see Fig. 7.1b) at $\lambda = 422$ nm is due to the excitonic transition of the dye. Being PVP a more favorable host for the dye, its presence allows to keep the concen-

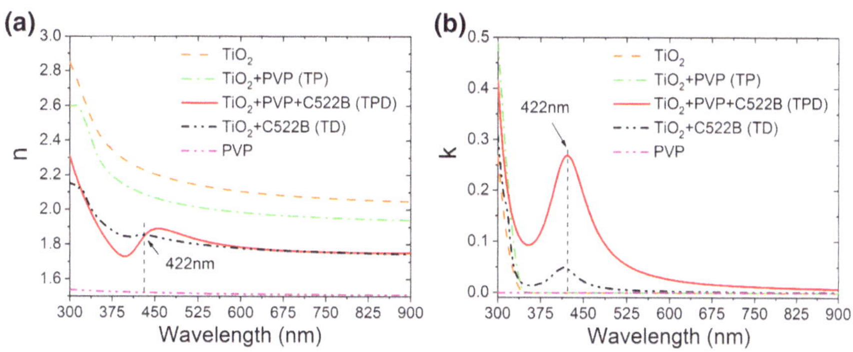

Fig. 7.1 Ellipsometric results for the real (**a**) and imaginary (**b**) refractive indices for pure sintered TiO_2 (dashed orange curve), TP (dot-dashed green curve), TPD (solid red curve), TD (dash-dot-dot black curve), and pure PVP (dash-dot-dot purple curve). Reproduced with permission from Ref. [5] Wiley

tration of dye in the resulting films suitably high to manifest acceptable photophysical properties. As a slight variation from the canonical sol-gel technique, the final TD and TPD samples did not undergo a final sintering procedure, in order to preserve the photoluminescence of the dye.

The effects of the heating process on the blend have been investigated by Variable Temperature Ellipsometry (VTE) experiments. Such a characterization involves the use of a customized heating stage to be placed on the goniometric sample holder of the Ellipsometer. As expected, a contraction of about ¼ of the initial thickness has been measured via VTE in the case of 80 nm thick unsintered TiO_2 layers, after one heating cycle up to a temperature of 80 °C.

This contraction endows the final HMM with thermo-chromic response. This effect is reversible within the typical application range of common electronic devices, from room temperature (RT) up to 80 °C. An analogous behavior is observed in the TPD layer being, at room temperature, about 75 nm thick. A yellow to orange color change is gradually observed while heating the TPD layer, outlining the role of the unsintered titania as a thermally reconfigurable template. Indeed, as shown in the inset of Fig. 7.2b the thermos-chromic response occurs only in presence of TiO_2.

To explain the origin of this phenomenon, a VTE investigation of the optical properties of the TPD was carried out, revealing that when heated up to 80 °C, its refractive index increases toward the values of sintered TiO_2 (compare Fig. 7.1a with Fig. 7.2a). The increase of temperature induces optical and morphological variations in the TPD layer, triggering a noticeable solvato-chromic thermo-responsivity of the embedded dye. As a consequence, the Lorentzian absorption peak present in the TPD layer red-shifts by about 45 nm between room temperature and $T = 80$ °C, as shown in Fig. 7.2a, b.

Thermo-gravimetric analysis (TGA), reported in the last part of the chapter, confirms that the temperature-induced reversible contraction of the layer responsible for the color change is due to water absorption/desorption from the environment.

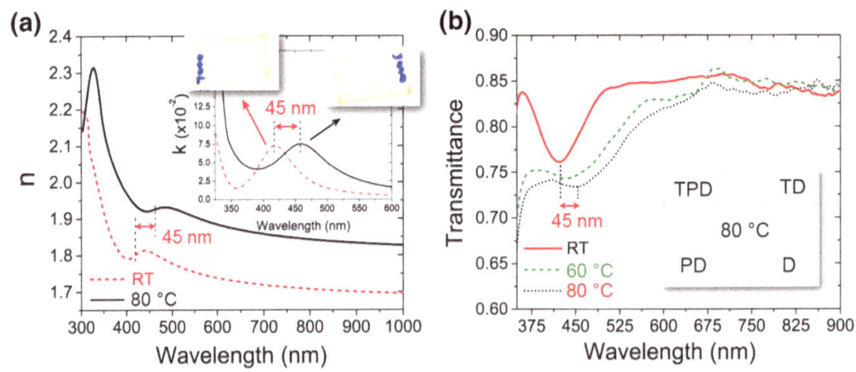

Fig. 7.2 **a** n and k (inset) of the TPD measured at room temperature (red-dashed curve) and at 80 °C (black curve). **b** Transmittance of TPD at different temperatures, illustrating a 45 nm thermal shifting between RT and 80 °C. Reproduced with permission from Ref. [5] Wiley

The photophysical properties of TPD were investigated by means of fluorescence spectroscopy and time-correlated single photon counting (TCSPC) measurements, illustrated in Fig. 7.3. The emission wavelength results red-shifted with respect to both PD (8 nm) and TD (1 nm): this points out the role of TiO_2 in the photoluminescence of the dye embedded in TPD. TCSPC measurements (Fig. 7.3b) show that while the PD solution (black curve) and film (red curve) decays lie in the same time range, the ones obtained from the TD (dashed-red curve) and TPD (dotted-red curve) films are significantly faster. The details are reported in Table 7.1.

In particular, in the PD solution a single lifetime τ_2 of about 4.83 ns has been extracted, while in the PD films two lifetimes are obtained: a longer one of 4.57 ns, which can be attributed to isolated dye molecules, and shorter one of 1.8 ns, attributed to aggregated molecules. In both the TD and TPD films, two faster decay times were evidenced, $\tau_{1,TD} = 0.46$ ns, $\tau_{1,TPD} = 0.31$ ns, $\tau_{2,TD} = 1.89$ ns, and $\tau_{2,TPD} = 1.65$ ns. The very short τ_1 observed in both TD and TPD can be attributed to an energy transfer process to the TiO_2 [5, 6]. Steady-state emission spectra were recorded by a HORIBA Jobin-Yvon Fluorolog-3 FL3-211 spectrometer equipped with a 450 W xenon arc lamp, double-grating excitation and single-grating emission monochromators and a Hamamatsu R928 photomultiplier tube. Time-resolved measurements were performed using the time-correlated single-photon counting (TCSPC) option on the Fluorolog-3. A NanoLED at 379 nm, full width at half maximum (FWHM)

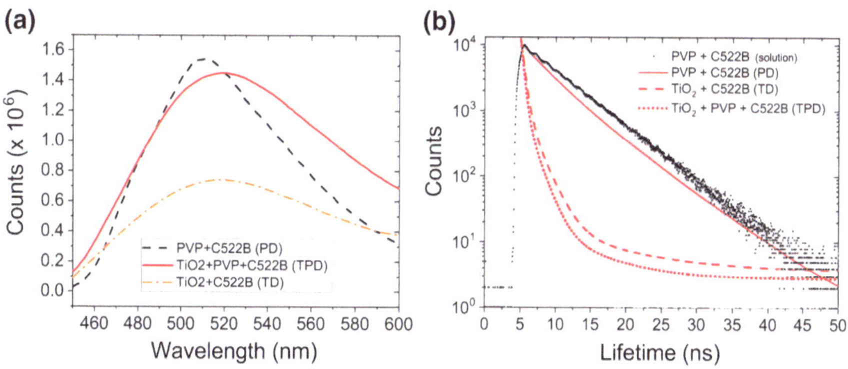

Fig. 7.3 **a** Photoluminescence for the three blends PD, TD, and TPD, together with **b** their corresponding decay lifetimes. Reproduced with permission from Ref. [5] Wiley

Table 7.1 Photophysical data of examined samples

Sample	λ_{em} (nm)	τ_2 (ns) (α/%)	τ_1 (ns) (α/%)
PVP + C522B (solution)	512	4.83	
PVP + C522B (PD)	511	4.57 (56.43)	1.8 (43.57)
TiO_2 + C522B (TD)	518	1.89 (29.62)	0.46 (70.38)
PVP + TiO_2 + C522B (TPD)	519	1.65 (27.18)	0.31 (72.82)

Reproduced with permission from Ref. [5] Wiley

< 200 ps with repetition rate at 1 MHz, was used to excite the samples. The excitation source was mounted directly on the sample chamber at 90° to a single-grating emission monochromator and collected with a TBX-04-D single-photon-counting detector.

7.3 Design, Fabrication and Characterization of the HMM Embedding the Thermo-Responsive Blend

A plasmonic hyperlattice constituted of five Ag/TPD (20/75 nm) bilayers has been realized. The dispersion relation for such a system is reported in Chap. 5 [Eq. (5.14)]. The topological transition from the effective dielectric to the *type II* anisotropy as expressed in Chap. 5, manifests as the occurrence of an epsilon-near-zero wavelength in $\varepsilon_{||}$, which results in a non-integrable singularity [6], easily calculable via EMT. Here it is important saying that the full HMM fabrication process requires the TPD layer to be placed under vacuum for the sputtering of the Ag layer. During the first vacuum process required by the sputtering technique, the TPD degasses significantly and a contraction of about 15 nm is induced (measured via ellipsometry). Such a contraction has to be considered in the calculation of the ENZ wavelength, since it modifies the final size of the TPD that turns to be about 60 nm. Spectroscopic ellipsometry allows also a fine evaluation of the complex refractive indices of both TPD and Ag. This reveals an ENZ condition for $\lambda = 502$ nm, where an effective-dielectric/*type II* transition occurs (see Fig. 7.4a). Due to the ENZ nature of this transition, light propagating through the HMM at this wavelength is expected to be extremely confined inside the dielectric layers.

The nature of this phenomenon is still under vivid investigation. It has been recently suggested that due the near-zero effective permittivity of the HMM at this wavelength results in Ferrel-Berreman like resonances [7]. It is still under debate whether these modes can be assimilated to artificial volume plasmons and if, as such, they can be excited via classic light or not [8, 9]. It is however true that the particular ENZ resonance considered in this chapter, occurring from the specific geometrical arrangement of the multilayer, is excitable by classic *p*-polarized plane wave, without any need for momentum matching techniques. A dip in *p*-polarized reflectance, experimentally detected by Ellipsometry at an angle of incidence of 40° confirms the ENZ nature of this mode, being also the signature of the topological effective dielectric/*type II* transition, as shown in Fig. 7.4b.

As expected, the system manifests very low reflectivity below λ_{trans}, while it becomes highly reflective above λ_{trans}, approaching the typical features of *type II* HMM. Figure 7.4b reveals also the presence of the Ferrell-Berreman mode, appearing as a dip in reflection around 325 nm. The Ferrell-Berreman mode is a particular case of radiative bulk plasmon polariton occurring at a metal/dielectric interface as well as in layered metal/dielectric subwavelength stacks. Such modes are of particular interest since they can be excited by free space propagating waves and occur at the ENZ

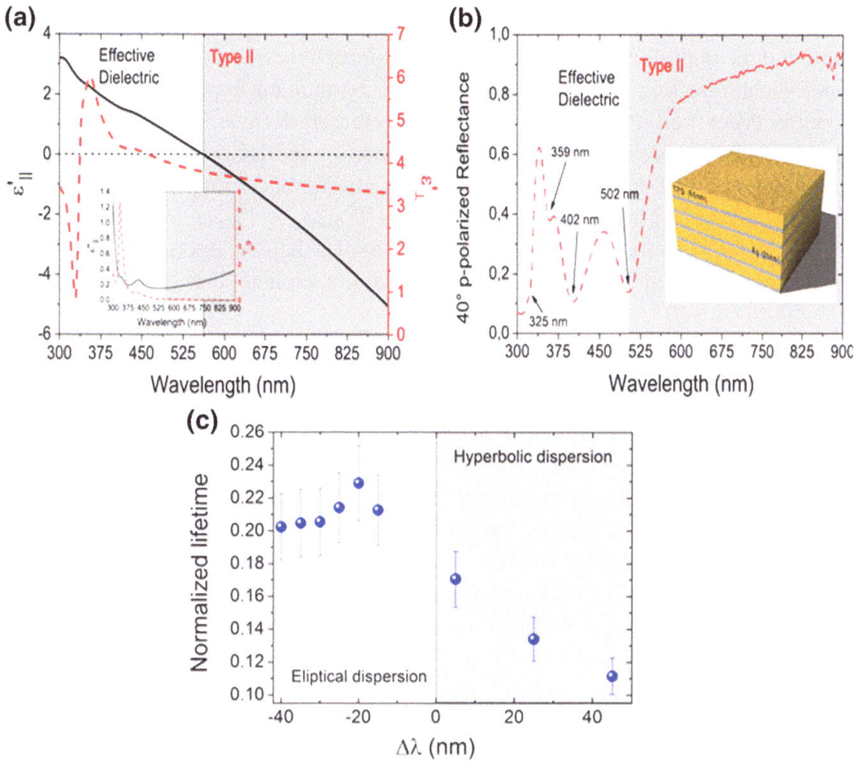

Fig. 7.4 **a** EMT design of the real and imaginary effective dielectric permittivities of the complete HMM. **b** Experimental p-polarized reflectance of 5 bilayer HMM at 40°. **c** Measured decay lifetimes in the HMM, normalized to a reference sample and plotted as a function of the difference between emission wavelength and transition wavelength. The decay rate is extremely increased within the *type II* hyperbolic region, in contrast to the flat trend shown in the effective dielectric (elliptical) range. Reproduced with permission from Ref. [5] Wiley

frequency of Ag itself. Two additional modes are present in the effective dielectric region, manifesting as reflection dips at 359 and 402 nm. Even though clarifying the nature of these modes is not the aim of this chapter, they can be seen as artificial, non-ideal volume plasmons. Avrutsky et al. [10] suggested that these modes are bulk-plasmon-polaritons, arising from the mutual interaction and hybridization of surface plasmons propagating along the metal/dielectric interfaces. Interestingly, a similar behavior has been described in multilayered spherical nanoparticle systems, being extremely useful in applications like thermo-plasmonic cancer treatment [11–13].

In Chap. 5, it has been reported how the hyperbolic dispersion typical of HMMs leads to the occurrence of high-k modes. These modes can modify the density of photonic states in proximity of the HMM and provide additional relaxation channels for fluorophores placed close to it. The weak interaction between the fluorophore emitting in the *type II* region and the HMM leads to noticeable Purcell effect with consequent significant modification of its decay lifetime. Such a feature is particularly evident if the photoluminescence (PL) of the fluorophore lies on the horseback of the topological transition, since the range of the PL lying in the hyperbolic region, experiencing Purcell effect, manifests strongly decreased decays. Indeed, in proximity of plasmonic cavities such as HMMs, the expression for decay rate of a fluorophore should be modified, taking into account the contribution coming from surface plasmons coupling and *high-k* modes, so that $\Gamma_{vac} + \Gamma_{plasmon} + \Gamma_{high\text{-}k}$. When the distance "*d*" of the fluorophore from the HMM is far smaller than the operative wavelength, the decay rate is dominated by the high-*k* relaxation channel that is equal to [14],

$$\Gamma_{\text{high}-k} \approx \frac{\mu^2 \text{Im}(r_p)}{8\hbar d^3} \tag{7.1}$$

Here r_p is the *p*-polarized reflectance of the HMM at the wavelength of interest and μ is the dipole moment. Approaching the *type II* anisotropy, the value of $\text{Im}(r_p)$, which is also known as topological parameter, starts increasing, so that $\Gamma_{high\text{-}k}$ plays a predominant role in increasing the decay rate of a fluorophore placed in proximity of the HMM. This behavior has been used as the technique of election for the fine positioning of the topological transition in several systems.

Time Correlated Single Photon Counting (TCSPC) measurements conducted on the HMM allowed to detect and position its topological transition exactly where predicted in the framework of EMT and in correspondence of the ENZ wavelength. Figure 7.4c shows that the normalized fluorescence lifetime (τ_1) of the dye embedded in the HMM (as a function of the emission wavelength reported as the distance from the transition wavelength exhibits a flat trend within the effective-dielectric range, while it is remarkably reduced within the type II region.

7.4 Thermal Tunability of the Optical and Photophysical Response of the HMM

The thermal tunability of the HMM was investigated by VTE experiments. When heated, the TPD dielectric layer undergoes water desorption, inducing a contraction and a relative increase of refractive index. As said before, during the first vacuum session, the TPD degasses, undergoing a consistent shrinking. After this, the compression induced by external factors cannot be substantial and, as such, the modification of the TPD refractive index is marginal. Consequently, once in the HMM, the temperature dependence of the refractive index of TPD is very low and the main role in the topological transition blue shift is played by the 10 nm thickness shrinkage. As shown in Fig. 7.5a, the reflectance response, and therefore the ENZ wavelength (Fig. 7.5b), undergoes a 25 nm blue-shift, upon heating from RT to T = 80 °C, which corresponds to a contraction of about 10 nm of the TPD layer.

Since the thickness shrinking and the heat-induced refractive index increase have opposite effects on the HMM topological transition shift, it has been possible to separately quantify each of their contributions by means of EMT calculations. The refractive index increases of the TPD layer as a consequence of the vacuum cycle needed for the sputtering deposition of Ag has been ellipsometrically evaluated, as shown in Fig. 7.5c. As a consequence of the first vacuum cycle, the TPD's refractive index is increased towards the one of the 80 °C free space heated TPD.

This effect makes so that the heat-induced refractive index increases of the TPD in the HMM is very low. Therefore, the main contribution to the topological transition shift is provided by the thickness shrinking, determining an overall blue-shift. In order to quantify and separate the effects due to, respectively, the heat-induced refractive index increase and the thickness shrinking, we calculated the topological transition

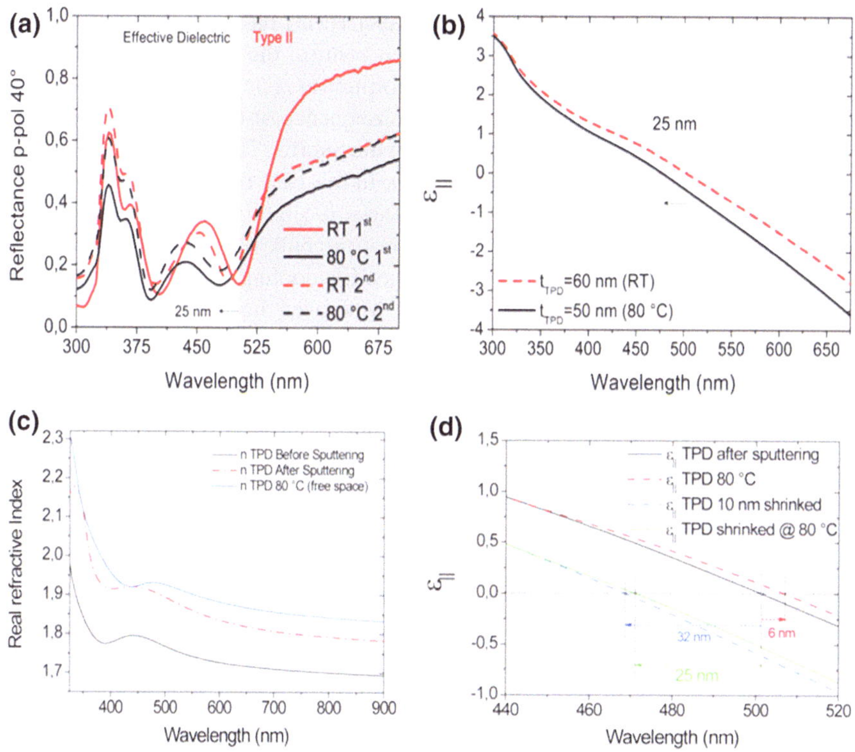

Fig. 7.5 **a** Experimentally detected reversible thermal tuning of the reflection response of the HMM. **b** Calculated EMT thermally induced blue shift of the HMM ε_{NZ} transition. **c** Refractive Index change of the TPD layer at room temperature before vacuum treatment (black curve), after vacuum treatment (red, dot-dashed curve) and at 80 °C (blue curve). EMT calculation of the effect of the heat induced refractive index increase (red, dashed curve), of the 10 nm thickness shrinking (blue, dashed curve) and of both of them (real case, green solid curve), with respect to the room temperature case (black, solid curve). Reproduced with permission from Ref. [5] Wiley

in the case of: (i) a hypothetical TPD layer with exactly the same thickness of the post-vacuum original TPD but with the 80 °C refractive index (this allows us to consider only the effect of the refractive index increase) and (ii) a hypothetical TPD layer with exactly the same refractive index of the post-vacuum original TPD, but with a 10 nm thinner layer (this allows to isolate the effect of the thickness shrinking). Results plotted in Fig. 7.5d show that in the former case, where only the increase of refractive index is present (red, dashed curve), a red shift of about 6 nm is found. On the other hand, when only the effect of the 10 nm shrinkage is present (blue, dotted curve), a blue shift of about 32 nm is induced. The neat result of the superposition of these two effects, that is the real case in which the shrinking and the refractive index increase are simultaneously present, is a blue shift of about 25 nm.

When the temperature is restored to the room level, the TPD layer reabsorbs from the naturally humid environment a consistent part of the lost water, leading the tunable response to be reversible (Fig. 7.6a, b). In order to further confirm the reversibility of the process, the spectral response of 3-bilayer HMM has been investigated after several heating/cooling cycles. Results, which confirm the perfect recovery of the topological transition apart from the initial desorption of residual solvent, are reported in Fig. 7.6a,b. It is demonstrated that, as a consequence of the heating process, the spectral position of the bulk plasmon polariton blue-shifts. This is directly ascribable to a thickness reduction of the dielectric layer. In this case the blue-shift is of about 10 nm, due to the fact that the TPD's thickness is slightly lower. However, it is important to notice that the spectral position of the bulk plasmon polariton of the 3-bilayers structure is preserved after three heating/cooling cycles, confirming the reversibility of the process after several cycles. Moreover, the inset of Fig. 7.6b shows that the presence and the quality of the Ferrel-Berreman mode is not affected by the heating/cooling cycles, demonstrating that Ag is not significantly oxidizing.

One important consequence of the heating cycle is the lowering of the reflectivity occurring in the hyperbolic region. The reason of this has to be searched in the roughening of the TPD surface after the heating process. When this layer is sandwiched between two Ag claddings, the upper metaling film tends to follow this roughening inducing a remarkable increase of the scattering. Unfortunately, a complete recovery of the original reflectivity is impossible, but the hyperbolic dispersion is not affected by this, since the enhanced corrugation does not prevent Ag to manifest its typical plasmonic features such as the presence of the Ferrell-Berreman mode, detected at 325 nm for p-polarized reflectance at an angle, as shown in Fig. 7.6b.

A deeper inspection of the same picture reveals that; however, the topological transition shift is not completely reversible. Indeed, when undergoing the first heating cycle, the TPD layer loses all the residual solvents, so that the first blue shift results slightly broader than the successive. The Thermo-Gravimetric Analysis (TGA, performed by means of a Perkin-Elmer Thermogravimetric Analyzer Pyris 6 TGA) showed in Fig. 7.6c conducted on the TPD layer confirms this hypothesis. Only 90% of the starting weight of the TPD solution is recovered after the first heating/cooling cycle. Interestingly, from the second TGA heating/cooling cycle on, almost the 100% of the starting weight is recovered, confirming the hypothesis formulated before.

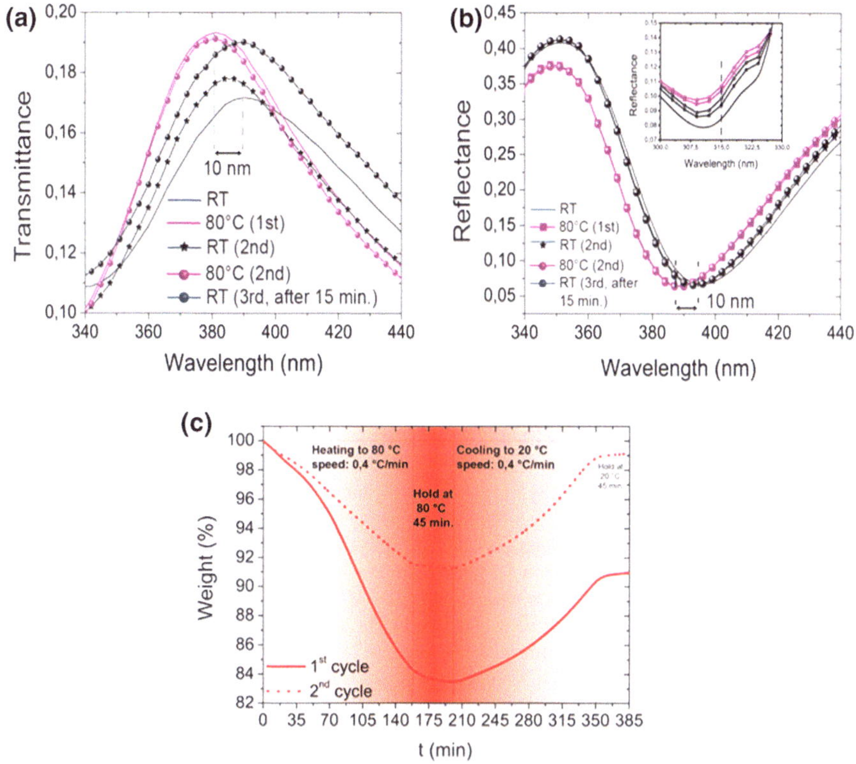

Fig. 7.6 **a** Transmission blue shift of the Bulk Plasmon Polariton wavelength of a new 3 bilayers structure. The spectral position of the BPP mode is restored after several heating/cooling cycles. **b** Reflectance behavior of the same sample. **c** TGA analysis conducted on the TPD layer. After the first heating/cooling cycle (blue solid line), TPD recovers only 90% of its original weight. The second cycle (dashed curve) shows, instead, an almost complete recovering of the starting weight. Data are normalized to the weight at the beginning of each heating cycle. Reproduced with permission from Ref. [5] Wiley

In the end, it is important to underline that the water absorption/desorption from humid environment can occur independently from an increase of temperature. As a consequence, the TPD can vary its optical characteristics also in response to a change in the dryness of the environment in which is immersed. To confirm this, the TPD has been positioned in dry atmosphere (water content below 1 ppm) and then removed, showing also in this case a remarkable color change.

References

1. De Sio L, Caligiuri V, Umeton C (2014) Tuneable broadband optical filter based on soft-composite materials. J Opt 16:065703
2. Caligiuri V, De Sio L, Petti L, Capasso R, Rippa M, Maglione MG, Tabiryan N, Umeton C (2014) Electro-/all-optical light extraction in gold photonic quasi-crystals layered with photo-sensitive liquid crystals. Adv Opt Mater 2:950–955
3. Papadakis GT, Atwater HA (2015) Field-effect induced tunability in hyperbolic metamaterials. Phys Rev B 92:184101
4. Caligiuri V, De Luca A (2016) Metal-semiconductor-oxide extreme hyperbolic metamaterials for selectable canalization wavelength. J Phys D Appl Phys 49:08LT01
5. Caligiuri V, Lento R, Ricciardi L, Termine R, La Deda M, Siprova S, Golemme A, De Luca A (2018) Environmental control of the topological transition in metal/photoemissive-blend metamaterials. Adv Opt Mater 6:1–8
6. Krishnamoorthy HNS, Jacob Z, Narimanov E, Kretzschmar I, Menon VM (2012) Topological transitions in metamaterials. Science 336:205–209
7. Caligiuri V, Palei M, Imran M, Manna L, Krahne R (2018) Planar double-epsilon-near-zero cavities for spontaneous emission and Purcell effect enhancement. ACS Photonics 5:2287–2294
8. Ciattoni A, Marini A, Rizza C, Scalora M, Biancalana F (2013) Polariton excitation in epsilon-near-zero slabs: transient trapping of slow light. Phys Rev A 87:1–8
9. Newman W, Cortes CL, Atkinson J, Pramanik S, DeCorby RG, Jacob Z (2014) Ferrell–Berreman modes in plasmonic epsilon-near-zero media. ACS Photonics 2:2–7
10. Avrutsky I, Salakhutdinov I, Elser J, Podolskiy V (2007) Highly confined optical modes in nanoscale metal-dielectric multilayers. Phys Rev B 75:241402–241406
11. Prodan E, Radloff C, Halas NJ, Nordlander PA (2003) Hybridization model for the plasmon response of complex nanostructures. Science 302:419–422
12. Prodan E, Nordlander P (2003) Structural tunability of the plasmon resonances in metallic nanoshells. Nano Lett 3:543–547
13. Ayala-orozco C, Urban C, Knight MW, Urban AS, Neumann O, Bishnoi SW, Mukherjee S, Goodman AM, Charron H, Mitchell T (2014) Au nanomatryoshkas as efficient transducers for cancer treatment: benchmarking against nanoshells. ACS Nano 8:6372–6381
14. Jacob Z, Smolyaninov II, Narimanov EE (2012) Broadband Purcell effect: radiative decay engineering with metamaterials. Appl Phys Lett 100:181105–181109

Chapter 8
Guided Modes of Hyperbolic Metamaterial and Their Applications

As mentioned in the previous chapters, HMMs support highly confined non-radiative modes such as *high-k* modes, in addition to surface plasmon modes within the structure due to hyperbolic dispersion. On the other hand, the *high-k* modes are conventionally referred to as volume plasmon polaritons (VPPs) or bulk Bloch plasmon polaritons (BPPs). It is a challenging task to excite and collect BPP modes especially at optical frequencies, since they are highly confined within the HMM. In this chapter, we focus on the excitation and collection of BPP modes of multilayered HMMs in the optical frequencies, using grating and prism coupling techniques. We further demonstrate that the proposed grating and prism coupled HMMs exhibit an extraordinary potential to open new routes towards a wide range of breakthrough applications such as biosensing, spontaneous emission enhancement, multiband perfect absorption and reconfigurable photonic devices.

8.1 Guided Modes of Hyperbolic Metamaterials

Hyperbolic metamaterials support radiative modes such as Ferrell-Berreman modes [1] and Brewster modes [2], which can be excited from the free-space. More importantly, HMM supports non-radiative modes such as surface plasmon polaritons (SPPs) and highly confined bulk plasmon polaritons in the hyperbolic dispersion [3]. In fact, BPPs are the entire family gap plasmon modes of a multilayered metal-dielectric stack and they are propagating wave inside the multilayer due to the coupling of individual gap plasmon modes. However, it is an exponentially decaying field outside the structure like SPPs. The modal indices of BPP modes vary from positive to negative, which are larger than that of SPPs. It has been theoretically and experimentally investigated the existence of BPP modes in HMMs [3–5]. Avrutsky et al. [3] experimentally verified their existence in a multilayered silicon-gold stack at 1550 nm wavelength using a prism coupling scheme, where they theoretically demonstrated that the BPPs represent the entire family gap plasmon modes of a metal-dielectric stack with symmetric field distribution. On the other hand, Zhukovsky et al. theoret-

Sreekanth K. V. et al., *New Directions in Thin Film Nanophotonics*, Progress in Optical Science and Photonics 6, https://doi.org/10.1007/978-981-13-8891-0_8

ically showed that the coupling of short-range surface plasmon polaritons (SRSPPs) is responsible for the existence of BPP modes in an HMM [5]. However, in the optical frequencies, the excitation of BPP modes of an HMM is a critical task due to their large modal indices and deep sub-wavelength confinement.

Even though the excitation of BPP modes of a multilayered *type II* HMM is experimentally demonstrated at near infrared frequencies by prism coupling principle, it is not possible to excite BPP modes in the visible spectral region using prism coupling due to the unavailability of high refractive index prisms that are transparent in the visible. In this context, the grating coupling principle could be an alternative solution to excite the *high-k* modes of an HMM. It has been experimentally demonstrated the generation of subwavelength interference patterns by integrating a subwavelength diffraction grating on top of a hyperbolic medium [6]. In this chapter, we experimentally demonstrate that it is possible to excite the *high-k* BPP modes of *type II* HMM at visible and NIR frequencies using grating coupling principle and *low-k* BPP modes of *type I* HMM at visible frequencies using prism coupling principle. In particular, the excitation of higher order BPP modes are demonstrated by integrating subwavelength metallic diffraction gratings with *type II* HMMs. In the proposed grating coupled HMMs, nanofabrication and nanophotonics strongly interplayed during the design and fabrication processes to create subwavelength layer thick HMM superlattices, and to implement a proper grating to couple and simultaneously probe SPP and BPP modes.

8.2 Excitation of Guided Modes of HMM

8.2.1 *Using Grating Coupling Technique*

The basic principle of grating coupling technique is that the surface plasmons in a sub-wavelength grating can be excited when the parallel wavevector of the surface plasmons are comparable with the wavevector of the illumination light. At this condition, grating diffraction orders are no longer propagating waves, but they are evanescent field. This evanescent field has enhanced wavevector, which is responsible for the coupling of incident light to the surface plasmon modes. The coupling condition is given by,

$$k_{spp}^2 = n_0^2 k_0^2 \sin^2\theta \pm 2n_0 m k_g k_0 \sin\theta \cos\phi + \left(mk_g\right)^2 \tag{8.1}$$

where θ being the incidence grazing angle, ϕ is the azimuthal angle (the angle between the plane of incidence and the grating wavevector), n_0 is the refractive index of incident medium, $k_0 = 2\pi/\lambda$ is the vacuum wavevector, m is the grating diffraction order and $k_g = 2\pi/\Lambda$ is the grating wavevector with Λ being the grating period.

When $\phi = 0$, Eq. (8.1) reduces to

$$k_{spp} = n_0 k_0 \sin\theta \pm m k_g \tag{8.2}$$

In the case of HMMs, a diffraction grating on top of the HMM can diffract light and produce a wide range of wavevectors into the HMM. In particular, the generated wavevectors can couple through the surface modes due to the existence of impedance mismatch at the various openings, [7].

In order to demonstrate the excitation of fundamental and higher order BPP modes of HMM via grating coupling principle, we initially fabricated a HMM consisting of 12 alternating layers of gold (Au) and titanium dioxide (TiO_2) thin films, which is shown in Fig. 8.1a. In the fabrication process, TiO_2 and Au thin films are sequentially deposited on a clean glass substrate by means of RF sputtering and thermal evaporation techniques, respectively. The measured thickness of TiO_2 and Au is 32 and 16 nm, respectively. In Fig. 8.1b, we show the EMT-derived real parts of the uniaxial permittivity tensor components of the fabricated TiO_2–Au HMM, where the frequency dependent optical constants of Au are obtained from Drude model, $\varepsilon_m = 1 - \left(\frac{\omega_p^2}{\omega(\omega + i/\tau)}\right)$, with ω_p is the plasma frequency of Au, ω is the excitation frequency and τ is the relaxation time. The dielectric constant of TiO_2 is set to be a constant value of 7.3. We further confirm the behavior of EMT-derived dielectric tensor components by performing spectroscopic ellipsometry measurements. As can be seen, the fabricated HMM shows *type II* hyperbolic dispersion in the visible spectral band with dielectric permittivity components $\mathrm{Re}(\varepsilon_{//}) < 0$ and $\mathrm{Re}(\varepsilon_{\perp}) > 0$, when $\lambda \geq$ 548 nm. Also note that the imaginary parts of both components ($\mathrm{Im}(\varepsilon_{//})$, $\mathrm{Im}(\varepsilon_{\perp}) > 0$) are positive throughout the spectral band (not shown).

As shown in Fig. 8.1c, we designed and fabricated a grating coupled-HMM (GC-HMM) to excite both the surface and bulk plasmon polaritons of the HMM. Here, we integrate a sub-wavelength metallic diffraction grating (1D or 2D) with a multilayered HMM [8, 9]. In the fabrication process, PMMA stripe-relief is initially patterned on top of the HMM using electron-beam lithography and then directly deposited 20 nm thick Ag film on the PMMA grating. In fact, we fabricated Ag diffraction grating because Ag has lower optical loss level in the visible compared to other metals and therefore, it has higher plasmonic coupling efficiency. In addition, we ensured that the over layer contamination is negligible while evaporating Ag directly on the PMMA stripe-relief, which is clear from the SEM image of as fabricated 1D Ag diffraction grating shown in Fig. 8.1d. It is visible that Ag is uniformly distributed on the PMMA grating. By using this simple procedure, a total area of about 5 mm × 5 mm is successfully patterned on the HMM, with fabricated gratings have an average period of 500 nm, average slit width of 160 nm and height of 120 nm. In order to avoid the direct contact between the grating and the HMM, and thus to improve the coupling efficacy, a thin TiO_2 spacer layer with thickness 10 nm is deposited between them.

To experimentally demonstrate the occurrence of both SPPs and BPPs in HMM, six different samples have been fabricated and the reflectance spectrum of these samples as a function of wavelength are measured using variable angle high-resolution ellipsometer, which is shown in Fig. 8.2. In our measurements, the polarization of

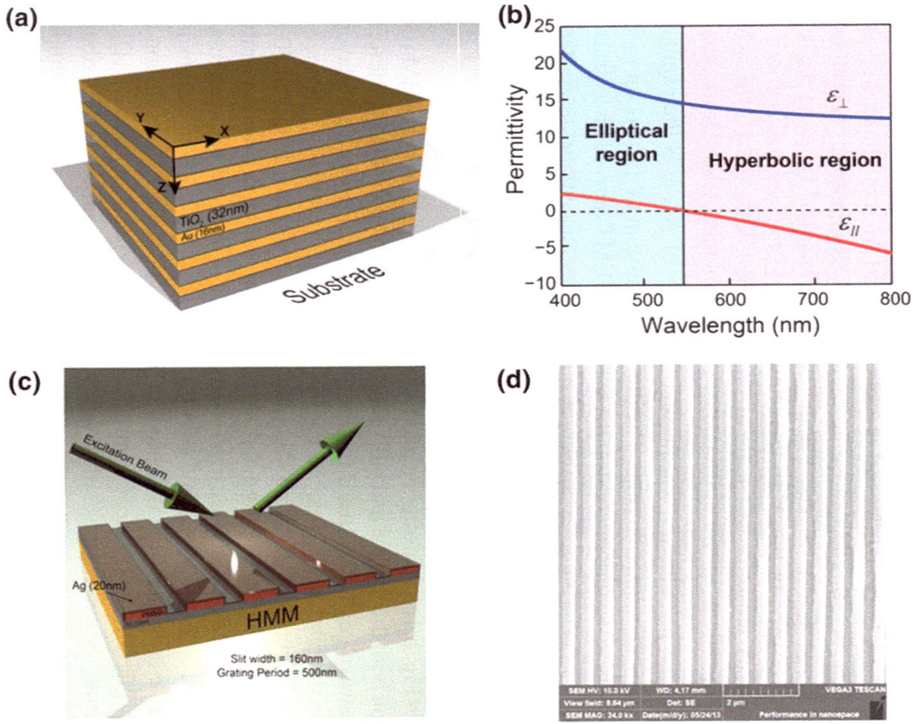

Fig. 8.1 **a** A schematic representation of the fabricated 6 bilayers of Au/TiO_2 HMM, **b** EMT-derived real parts of the effective uniaxial permittivity components of HMM. **c** A schematic diagram of the excitation of BPP modes of GC-HMM in reflection geometry and **d** SEM image of fabricated sub-wavelength 1D Ag diffraction grating on top of the HMM with an average period of 500 nm and average slit width of 160 nm. Reproduced with permission from Ref. [8]

incident beam is set to p-polarization and the angle of incidence is fixed at 50°. We first obtained the reflectance spectrum of a reference sample, which is an Ag grating on TiO_2/glass substrate (Fig. 8.2a). Two reflectance minima are observed in the wavelength range from 350 to 450 nm, which represents the SPPs supported by the reference sample. The acquired reflectance spectrum of a control sample (Ag grating on a single bilayer of Au/TiO_2) is shown in Fig. 8.2b. In contrast to Fig. 8.2a, the control sample shows three extra reflectance minima at wavelengths such as 700, 1000 and 2000 nm, in addition to other two reflectance minima of SPPs. These resonances at higher wavelengths represent the gap plasmon (or bulk plasmon) modes of control sample. Further, we obtained the reflectance spectrum in samples which consist of Ag grating on 4 bilayers of Au/TiO_2 and observed almost same behavior like a control sample (Fig. 8.2c). However, both SPP and BPP bands are slightly modified. Figure 8.2d shows the reflectance spectrum of the proposed GC-HMM. As can be seen, the reflectance minima at wavelengths such as 700, 1000 and 2000 nm are very narrow and deeper as compared to Fig. 8.2b, c. These deep and narrow modes in the reflectance spectrum represent the existence of highly confined bulk plasmon polaritons (super modes) in the GC-HMM. It is also clear that the resonance wavelengths are slightly shifted when the number of bilayers is increased from 1 to 6 due to the coupling of individual gap plasmon modes.

As shown in Fig. 8.7d, both surface and bulk plasmon modes blue shift when the incidence angle increases, which is attributed to the variation in modal indices of guided modes with incidence angles. Also note that the blue shift in BPP mode band increases with increasing wavelength. The dispersion relations of both SPP and BPP modes are calculated to validate the experimental results. Figure 8.3a show the dispersion plot of SPP and BPP modes, where the SPP and BPP dispersion is obtained from $k_{SPP} = k_0\sqrt{\varepsilon_d\varepsilon_m/(\varepsilon_d + \varepsilon_m)}$ and $k_{BPP_0} = k_0\sqrt{\frac{\varepsilon_d\varepsilon_m(t_d+t_m)}{t_d\varepsilon_m+t_m\varepsilon_d}}$, respectively. One can see that both dispersions are well beyond the air light line. It is important to note that the SPP wavevectors are much higher than that of BPP at shorter wavelengths ($\lambda < 450$ nm) and the wavevectors of BPP modes are higher than that of SPP when the excitation wavelength is increased from 450 nm. Therefore, it can be concluded that the experimentally observed resonances in reflectance spectrum at shorter wavelengths (350–450 nm) are due to SPPs of diffraction grating and at higher wavelengths (700, 1000 and 2000 nm) are due to BPP modes of HMM.

As shown above, the BPP modes of the HMM are excited by coupling the SPPs of diffraction grating with the SPPs at each metal/dielectric interfaces of the HMM. It can be further experimentally evidenced by studying the reflectance spectra as a function of incidence angle and comparing the results with their modal indices. In Fig. 8.4a–f, we illustrate the excitation of BPPs at different excitation wavelengths. Here, we considered three BPP bands such as 650–750 nm, 1000–1100 nm and 2000–2100 nm and compared the results with that of a control sample (Ag grating on a single bilayer of Au/TiO_2). In contrast to control sample, high quality factor resonance in reflectance spectrum of each band is obtained for GC-HMM. These results further evidence the excitation of BPPs from the GC-HMM. A large coupling angle shift is observed from control to GC-HMM for the longer wavelength BPP mode

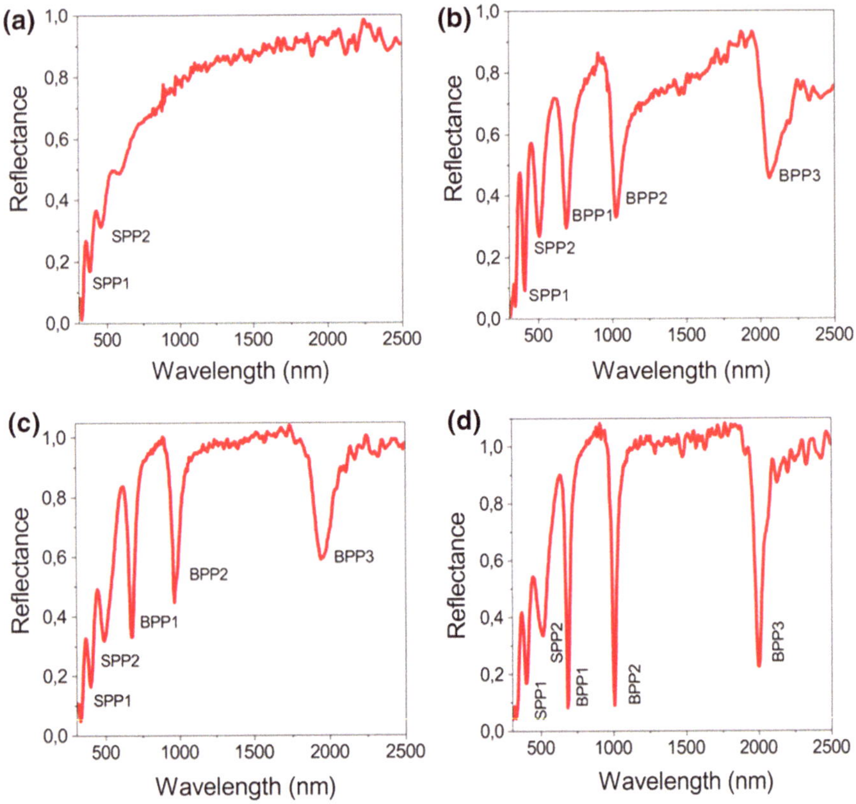

Fig. 8.2 Measured reflectance spectrum at 50° angle of incidence for **a** Ag grating on TiO_2/glass substrate, **b** Ag grating on 1 bilayer of Au/TiO_2, **c** Ag grating on 4 bilayers of Au/TiO_2 and **d** Ag grating on a 6 bilayer of Au/TiO_2. Reproduced with permission from Ref. [8]

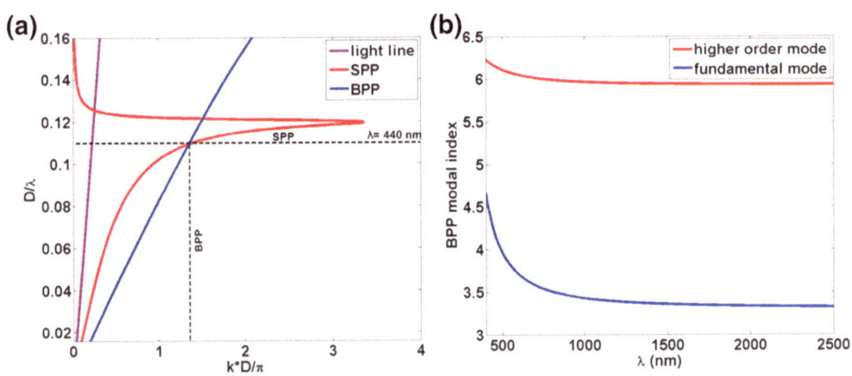

Fig. 8.3 **a** Dispersion relation of SPP and BPP modes. Both x and y axes are normalized with respect to period, D of the HMM. **b** BPP modal index variation with wavelength. Reproduced with permission from Ref. [8]

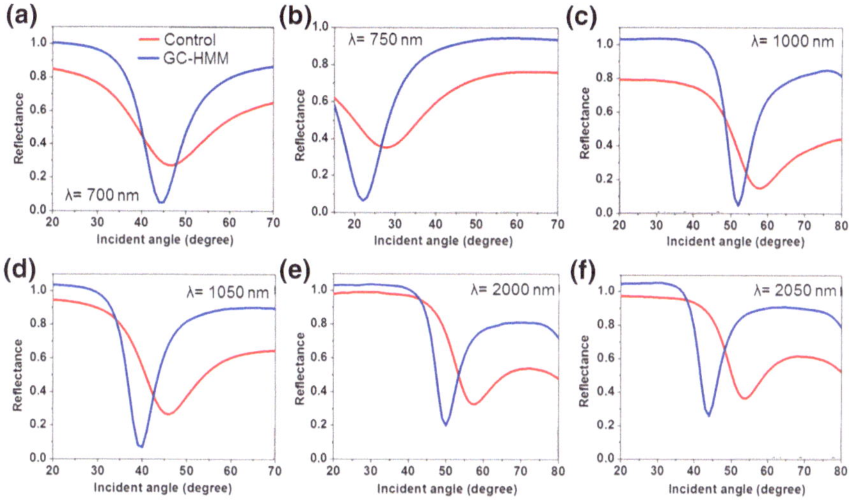

Fig. 8.4 Excitation of BPPs at different BPP mode bands for GC-HMM and control sample: **a** λ = 700 nm, **b** λ = 750 nm, **c** λ = 1000 nm, **d** λ = 1050 nm, **e** λ = 2000 nm and **f** λ = 2050 nm. Reproduced with permission from Ref. [8]

band, which is due to the strong mode confinement of BPPs at longer wavelengths. In addition, a decrease in coupling angle is obtained when the excitation wavelength is increased in each BPP mode band. This is attributed to the decrease in modal index of BPP modes with increasing excitation wavelength.

The modal index of the excited BPP mode can be experimentally obtained from grating coupling equation, $n_{\text{modal}} = ((\lambda/\Lambda) + n_0 \sin\theta)$ with λ is the excitation wavelength, Λ is the grating period and θ is the angle of incidence. The experimentally determined modal index of BPP modes in three BPP mode bands are tabulated in Table 8.1. The theoretically calculated fundamental and higher order modal index of BPP modes are shown in Fig. 8.3b. It is evident from the theoretical results that the modal index of BPP mode increases with decreasing the excitation wavelength and it slightly varies when the excitation wavelength increases. As shown in Table 8.1, we observed almost similar behavior in our experiments as the higher modal index is obtained for the shorter wavelength BPP band (650–750 nm) and it is almost constant at higher wavelength BPP band (2000–2100 nm). On the other hand, the observed large coupling angle shift in shorter wavelength BPP band is due to the large modal index variation of BPP modes. However, the coupling angle shift is minimum in longer wavelength BPP band due to almost constant modal index variation. It is also clear from the modal index analysis that the modal index of BPP mode increases with increasing the BPP mode wavelength band. Therefore, it can be concluded that the excited modes in the shorter wavelength BPP band (650–700 nm) represent the fundamental BPP modes and the excited modes in two longer wavelength bands represent the higher order modes. In short, the fundamental and higher order BPP modes of an HMM can be excited by properly coupling a sub-wavelength metallic diffraction grating with an HMM.

Table 8.1 Experimentally determined modal index of BPP modes

Wavelength (nm)	Coupling angle (°)	Modal index
Wavelength band: 650–750 nm (Fundamental modes)		
650	64	2.2
700	44	2.09
730	34	2.02
750	22	1.87
Wavelength band: 1000–1100 nm (First order modes)		
1000	52	2.8
1050	40	2.74
1080	32	2.7
1100	25	2.62
Wavelength band: 2000–2100 nm (Second order modes)		
2000	50	4.77
2030	46	4.78
2050	44	4.79
2100	38	4.81

8.2.2 Using Prism Coupling Technique

As mentioned in Sect. 8.1, the prism coupling scheme is not able to excite the high-k guided modes of multilayered HMMs in the visible spectral band due to the large effective index of HMMs. However, here we experimentally show that the low-k guided modes of a multilayered HMM can be excited using prism coupling technique [10]. In order to achieve this, we develop a multilayered *type I* HMM at visible frequencies since it has low-loss compared to *type II* HMM because of one negative permittivity component and it supports both low-k and high-k modes. To develop a low-loss *type I* HMM, ten alternating thin layers of a low-loss plasmonic material such as TiN and a low-loss phase change material such as Sb_2S_3 are deposited on a glass substrate (Fig. 8.5a). Since the dielectric layer of the HMM is a phase change material, the optical properties of the developed HMM can be tuned by switching the structural phase of Sb_2S_3 from amorphous to crystalline. The EMT-derived uniaxial permittivity components of a multilayered Sb_2S_3-TiN HMM when Sb_2S_3 in amorphous and crystalline phases are shown in Fig. 8.5b. It shows that HMM exhibits *type I* hyperbolic dispersion at $\lambda \geq 580$ nm when Sb_2S_3 is in amorphous phase, where $\varepsilon_{\perp} = \varepsilon_z < 0$ and $\varepsilon_{//} = \varepsilon_x = \varepsilon_y > 0$. However, the operating wavelength of *type 1* region is slightly blue shifted to 564 nm wavelength after crystallizing the Sb_2S_3 layers in the HMM.

A custom-built angular surface plasmon resonance (SPR) spectroscopy set-up is used for this purpose, where a right-angle BK7 prism is used in the Kretschmann configuration to match the momentum between the incident light and the HMM guided mode [10]. Transverse magnetic (TM) polarized light with a wavelength of

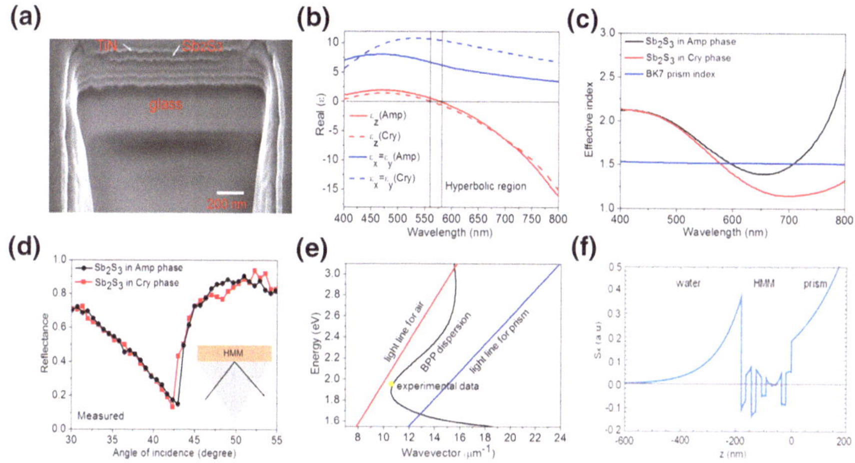

Fig. 8.5 **a** A cross-sectional SEM image of the fabricated Sb_2S_3-TiN HMM consisting of 5 pairs of Sb_2S_3 and TiN, with a thickness of 25 nm Sb_2S_3 and 16 nm TiN. **b** Real parts of uniaxial permittivity components of Sb_2S_3-TiN HMM when Sb_2S_3 is in the amorphous and crystalline phases. **c** Effective index of fundamental mode of Sb_2S_3-TiN HMM. **d** Experimentally obtained reflectance spectrum of Sb_2S_3-TiN HMM for both phases of Sb_2S_3. **e** Dispersion diagram for fundamental BPP mode of Sb_2S_3-TiN HMM. **f** Poynting vector of the guided mode of HMM. Reproduced with permission from Ref. [10] Wiley

632.8 nm is used as the excitation source because this wavelength belongs to the hyperbolic region of the HMM where its effective index is less than the prism index (Fig. 8.5c), so that the incoming beam has enough momentum to match the momentum of the HMM guided modes. Figure 8.5d shows the experimentally acquired reflectance spectrum of Sb_2S_3-TiN HMM when Sb_2S_3 is in both phases. Like typical reflectivity versus incidence angle curve of SPR, the excitation of the HMM guided mode is recognized as the coupling angle at which minimum reflected intensity occurs. We observed changes in the reflectance spectrum after switching the phase of the Sb_2S_3 from amorphous to crystalline state: (i) the minimum reflected intensity decreased, (ii) linewidth of the spectrum reduced and (iii) a decrease in coupling angle is obtained. These changes are due to the decrease in effective index of HMM with the phase change of Sb_2S_3. In Fig. 8.5e, we plot the dispersion diagram of fundamental BPP mode of Sb_2S_3-TiN HMM. It is important to note that the experimentally determined parallel wavevector at 632.8 nm wavelength is exactly on the BPP dispersion curve of HMM. The calculated Poynting vector (S_x) of the HMM guided mode at 632.8 nm is shown Fig. 8.5f. It indicates that the excited mode is a propagating wave inside HMM and decay exponentially at HMM-water (air) interface, and is leaky in the prism, which is the characteristic of a typical BPP mode of HMM. The propagating mode nearest in momentum at resonance is at 13.08 μm^{-1}, with a propagating length of 177 nm. Therefore, it can be concluded that the excited mode at 632.8 nm wavelength is the fundamental BPP mode of Sb_2S_3-TiN HMM, which is a low-k mode.

At the coupling angle of SPR reflectance spectrum, the phase difference between TM and TE polarized light experiences a sharp singularity [11]. By switching the phase of the HMM from amorphous to crystalline, this phase difference can be actively tuned, which is shown in Fig. 8.6a. This tunable phase difference is experimentally confirmed by varying the external dielectric constant [10]. Moreover, a tunable GH shift can be realized using Sb_2S_3-TiN HMM since the phase derivative at the coupling angle determines the magnitude of GH shift. Since we use prism coupling to excite the guided modes of the proposed HMM, it follows the same principle of a typical SPR system [12, 13] to enhance the GH shifts. That is, the guided mode of the HMM can be excited at the prism-HMM interface once the phase matching condition is satisfied through the prism, as a result a phase singularity can occurs at the coupling angle. Consequently, a large lateral shift between the incident and the reflected beam would generate at prism-HMM interface because the excited guided mode will propagate through the HMM and reflect out of the prism. One of the important advantages of the HMM in prism-coupled configuration to use in GH shift enhancement purpose is that the z-component wavevector (k_z) of the HMM can be modulated due to the unique anisotropic property of the HMM. Figure 8.6b represents the calculated GH shift of the prism-coupled HMM system and it is clear that the maximum GH shift is obtained at the coupling angle where there is a sharp change of phase difference. It is also shown that tunable GH shift is possible by switching the phase of the Sb_2S_3 in the HMM.

In order to demonstrate the tunable GH shift experimentally, we developed a reconfigurable sensor platform based on prism-coupled HMM system since the GH shift strongly depends on the superstrate dielectric constant. To record the tunable GH shifts, a home-built differential phase-sensitive set up is used [10]. We performed bulk refractive index sensing by injecting different weight ratios (1–10% w/v) of aqueous solutions of glycerol with known refractive indices into sensor channel. In Fig. 8.6c, we report the real-time tunable refractive index sensing by measuring the GH shift change with time due to the refractive index change of glycerol solutions. As can be seen, we realized a clear step function in GH shift by varying the glycerol

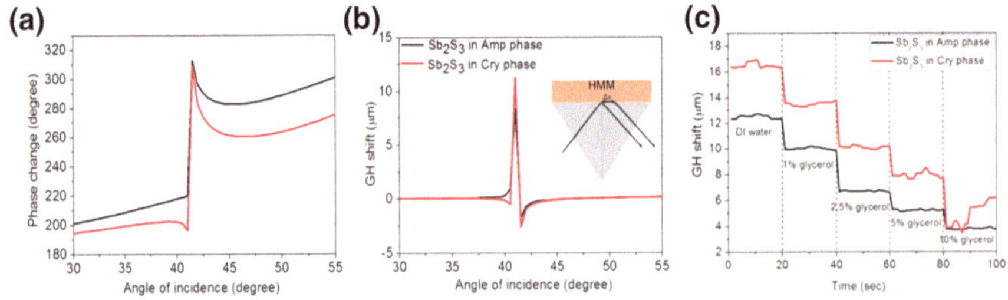

Fig. 8.6 For both phases of HMM in prism configuration **a** Calculated phase difference between TM and TE polarization, **b** Calculated GH shift. **c** Experimental demonstration of tunable GH shift. GH shift change with time by injecting different weight percentage concentrations of glycerol in distilled water. Reproduced with permission from Ref. [10] Wiley

concentration and a tunable GH shift by switching the HMM phase from amorphous to crystalline.

Since the crystalline phase provides higher phase change at the coupling angle compared to amorphous phase, we recorded maximum refractive index sensitivity for the crystalline phase of HMM, which is 13.4×10^{-7} RIU/nm, whereas a minimum sensitivity of 16.3×10^{-7} RIU/nm is recorded for the amorphous phase of HMM. It indicates that the sensitivity of the sensor can be tuned between these two values by switching the structural phase of the HMM via electrically or optically. Even though the obtained tunable range is small, it is possible to further improve the GH shift tunability by using longer wavelength sources and higher refractive index prisms [10]. This could be a promising opportunity to design an assay for the selective detection of higher- and lower-molecular-weight biomolecules by utilizing the tunable differential response of the HMM sensor. We further calculated the bulk refractive index sensitivity of the sensor, which is also of the order of 10^{-7} RIU/nm of the shift. By exploiting the GH shift enhanced refractive index sensitivity of the proposed HMM sensor, it is possible to detect small biomolecules at very low concentrations [10].

8.3 Applications of Grating-Coupled HMMs

8.3.1 Ultrasensitive Plasmonic Biosensing

Plasmonic biosensors are particularly valuable for the diagnosis of diseases and routine point-of-care (POC) clinical evaluations since it allows the rapid detection of biomolecular interactions in real-time [14]. The plasmonic biosensors work based on the principle of the excitation of surface plasmons at the metal/dielectric interfaces. As detailed in previous chapters, surface plasmons can be excited at the metal/dielectric interface when the wavevector of incident light satisfies the resonant conditions [15]. For visible and NIR frequencies, the electric field associated with these oscillations decay exponentially away from the metal/dielectric interfaces, which are highly sensitive to the small changes in refractive index of its surrounding medium. Therefore, plasmonic sensors facilitate label-free detection of biomolecules and real-time monitoring of biomolecular binding events.

In recent years, tremendous progress has been accomplished in micro- and nanofabrication technology, which has encouraged the development of high-performance label-free plasmonic biosensors based on metamaterials, which can overcome the limitations of conventional plasmonic biosensors [16–18]. In particular, a *type I* HMM consisting of 2D porous gold nanorod arrays has been developed for improving the sensitivity of plasmonic biosensors [19]. The authors have achieved a record bulk refractive index sensitivity of 30,000 nm/RIU with a maximum probe depth of 500 nm, by exciting the guided modes of plasmonic nanorod metamaterial via prism coupling technique. However, this system excites only one guided mode of the *type I* HMM at NIR frequencies. Thus, this sensor device showed enhanced sensitivity solely in the infrared spectral region and the sensing mechanism remains based on the bulky prism configuration. Even though commercial biosensors based on miniaturized prism configuration is available in the market, it is not suitable for point-of-care (POC) applications. In order to detect and monitor the real-time binding of small numbers of lower-molecular-weight (<500 Da) biomolecules such as proteins, hormones and drugs, sensors with higher figure-of-merit (FOM) and ultra-high sensitivity are required.

To address the above-mentioned limitations of plasmonic biosensors, a GC-HMM based multimode and miniaturized plasmonic biosensor platform is proposed [20–22], which shows broadband extreme sensitivity from visible to NIR wavelengths. In particular, we show that a multilayered *type II* HMM coupled to unconventional gold diffraction grating consisting of subwavelength periodic holes with square lattice symmetry can be used to realize high-sensitivity biosensors, which operates simultaneously at visible and NIR frequencies. An important advantage of GC-HMM geometry is that the resonance wavelength of the excited guided modes can be tuned from visible to NIR by changing both the HMM and the diffraction grating parameters. The spectral positions of the resonance can also be tuned by using a phase change material-based GC-HMM. More importantly, the practical realization of a miniaturized and multiplexed sensor device can be developed since the proposed configuration is based on a grating-coupling technique.

As shown in Sect. 8.2.1, the BPP modes of HMM show different spectral and angular features in each BPP mode band. Because of this distinct mode behavior, it is possible to design a potential multimode plasmonic biosensor with different spectral and angular sensitivities. Thus, we developed a miniaturized biosensor platform by integrating microfluidic channel with GC-HMM. The working principle of this sensor is based on the coupling condition between grating surface modes and BPP modes, which allows one to observe a change in resonance wavelength and resonance angle when the refractive index of the surrounding medium changes.

The proposed plasmonic biosensor platform, as illustrated in Fig. 8.7a, b, is a combination of a metallic diffraction grating and an HMM. Sixteen alternating thin films of gold and aluminum dioxide (Al_2O_3) with thickness 16 and 30 nm, respectively are deposited on a clean glass substrate to develop HMM which shows hyperbolic dispersion in the optical spectral range. According to Fig. 8.7c, the fabricated Au–Al_2O_3 multilayer is a *type II* HMM with hyperbolic dispersion at $\lambda \geq 520$ nm. To excite the *high-k* modes associated with Au–Al_2O_3 HMM via grating coupling technique, a 2D Au sub-wavelength diffraction is integrated with HMM (inset of Fig. 8.7a). In the fabrication process, a subwavelength hole array with an average period of 500 nm and hole diameter of 160 nm is patterned on top of the 10 nm thick Al_2O_3 spacer layer of the HMM using electron-beam lithography. Since Au is a most popular plasmonic material for biosensing applications due to its low oxidation rate and high biocompatibility, a 20 nm thick Au layer is directly deposited on top of the PMMA strip-relief to fabricate the required gold diffraction grating for coupling purpose.

By using a variable angle high-resolution spectroscopic ellipsometer, we acquired the reflectance spectra of the GC-HMM as a function of excitation wavelength and incidence angle to show the existence of BPP modes in the fabricated HMM. As

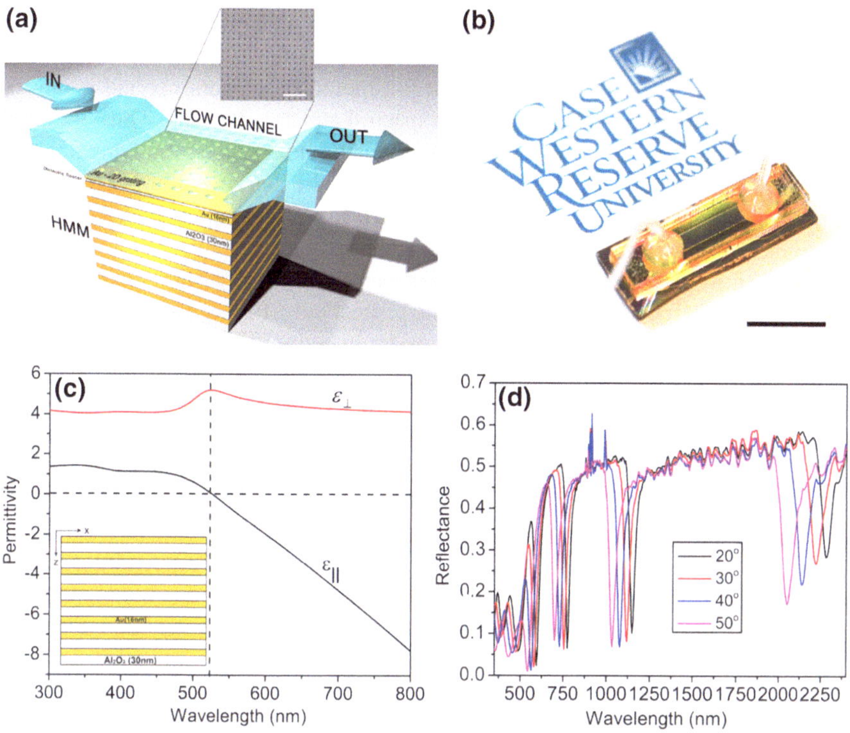

Fig. 8.7 **a** A schematic illustration of the fabricated miniaturized GC-HMM sensor device with a microfluidic channel and a SEM image of the patterned 2D subwavelength Au diffraction grating on top the HMM with an average period of 500 nm and hole size of 160 nm (Scale bar = 2 μm). **b** A photograph of the GC-HMM sensor device fully integrated with a microfluidic channel and sample tubing. Scale bar = 10 mm. **c** Real parts of effective permittivity of gold/Al_2O_3 HMM determined using effective medium theory, which shows a hyperbolic dispersion at $\lambda \geq 520$ nm. The fabricated eight pairs of gold/Al_2O_3 HMM is shown in the inset. **d** Reflectance spectra of the GC-HMM at different angles of incidence. Reproduced with permission from ref [20], Springer Nature

mentioned before, BPP modes represent the entire family of gap plasmon modes of a multilayer with large modal indices, which corresponds to strong mode confinement and shorter propagation length. Thus, BPP modes provide high quality (Q) factors. Figure 8.7d represents the reflectance spectrum of an Au grating coupled Au–Al_2O_3 HMM at different angles of incidence. The observed narrow modes above 500 nm wavelength represent highly confined fundamental and higher order BPP modes of Au–Al_2O_3 HMM. The obtained Q-factor of the modes at resonance wavelengths 1120, 755 and 580 nm are 29.5, 26 and 23, respectively. It indicates that the mode at 1120 nm provides maximum Q-factor and the mode at 580 nm provides minimum Q-factor, which could find possible applications in developing multi-analyte biosensors. However, we did not consider the longer wavelength mode at 2200 nm because the Q-factor of this mode is lower compared to other three BPP modes and, also this mode is not present in the reflectance spectrum after integrating the HMMs with a microfluidic channel.

It has already been demonstrated that plasmonic biosensors operate mainly in two major interrogation schemes such as spectral and angular scan. Thus, we study the performance of the GC-HMM sensors by using both spectral and angular scan interrogation schemes. As a first step to determine the detection limit of the sensor in both spectral and angular scan, we injected different weight ratios of aqueous solutions of glycerol into the sensor microchannel with a sample volume of $14 \times 2 \times 0.05\ mm^3$. The sensing parameters in our experiment is the wavelength and angular shifts in spectral and angular scan, respectively. Figure 8.8a, b respectively show the reflectance spectra of the sensor by injecting different weight ratios of glycerol in distilled water (0.1–0.5% w/v) in spectral and angular scan. It is clear that resonance wavelength corresponding to each BPP mode red shifted and the quality factor of each mode declined with increasing glycerol concentration. In addition, a positive angular shift is obtained at the four BPP mode excitation wavelengths when the glycerol weight ratio is increased (Fig. 8.8b). It shows the ability of the sensor to record extremely small refractive index changes of glycerol concentrations in both scans, for example a significant shift of 12 nm is obtained at 1300 nm even with 0.1% w/v glycerol solution. Interestingly, the shift in both spectral and angular scan increases when the spectral position of the BPP mode increases from visible to NIR wavelengths because the transverse decay of the field in the superstrate strongly varies from one mode to another [20]. In fact, the shifts vary nonlinearly with glycerol concentrations, which indicates that the sensor has its highest performance at the lowest concentrations.

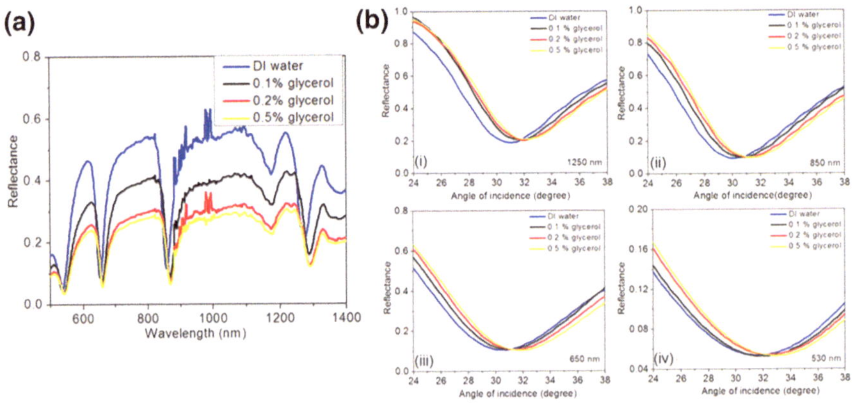

Fig. 8.8 Standard sensor calibration test in spectral interrogation scheme, by injecting different weight percentage concentrations of glycerol in distilled water. **a** Reflectance spectra of sensor device with different weight percentage of glycerol (0.1–0.5% w/v) and **b** Standard sensor calibration test in angular interrogation scheme. (i) at 1250 nm, (ii) at 850 nm, (iii) 650 nm, and (iv) 530 nm. Reproduced with permission from Ref. [22]

The spectral and angular detection limit of the sensor device can be determined from the obtained bulk refractive index sensing data. In particular, we determine the spectral and angular sensitivity of the device for each BPP mode band by using the marginal shift obtained for 0.5% (w/v) glycerol solution with respect to DI water. Since the refractive index of different concentrations of glycerol in DI water is well-known, the refractive index contrast between 0.5% (w/v) glycerol in DI water and DI water is around 0.0006. More importantly, the sensor provides different sensitivity from visible to NIR wavelengths because each BPP mode shows different shifts. The maximum spectral and angular sensitivity is obtained for longer wavelength BPP mode, which is around 30,000 nm/RIU and 2,500 degree/RIU, respectively, whereas the minimum sensitivity is recorded for shorter wavelength BPP mode, which is 13,333 nm/RIU and 2,333 degree/RIU for spectral and angular scan, respectively. The most important parameter of a sensor is the FOM, which is defined as $(\Delta\lambda/\Delta n)(1/\Delta\omega)$ with $\Delta\lambda$ being the wavelength shift, Δn being the refractive index change and $\Delta\omega$ being the full width of the resonant dip at half-maximum. By taking into account the sharpness of the resonance, the FOM determines the sensitivity with which very small wavelength changes can be measured. The developed sensor exhibits different FOM for each BPP mode, which are 206, 357, 535 and 590 at 550, 660, 880 and 1300 nm, respectively. It should be noted that the obtained FOM of the proposed *type II* HMM-based sensor is much higher as compared to existing plasmonic biosensors. For example, the previously developed *type I* HMM-based sensor exhibits only one guided mode with a FOM of 330 [19]. A promising advantage of GC-HMM sensor is the flexibility in the selection of a particular mode for the identification of specific biomolecules since it provides different sensitivities and FOM for each BPP mode. In particular, the differential response of the sensor device has the capability to use a lower sensitivity mode for the detection of higher molecular weight biomolecules and a higher sensitivity mode for the detection of lower molecular weight biomolecules.

Since higher FOM and ultra-high sensitivity are the two main requirements to detect smaller molecular-weight biomolecules (<500 Da) at lower concentrations, by using both spectral and angular interrogation schemes, we demonstrate the capabilities of the proposed sensor device for the detection and real-time binding of smaller numbers of biotin (molecular weight, 244 Da). The reason behind the selection of biotin protein is that it is a model system for small molecule compounds such as vitamins, cancer-specific proteins, hormones, therapeutics, or contaminants such as pesticides or toxins. More importantly, cells and proteins represent important blood-based biomarkers for non-invasive screening and, in most cases, require the detection of low cell numbers or ultralow concentrations of proteins from blood.

We consider longer wavelength mode to demonstrate biosensing because it showed maximum sensitivity and FOM. In order to capture biotin, we functionalized the sensor surface with streptavidin molecules. The sensor monitors the shifts due to the refractive index change caused by the capture of biotin at the sensor surface. The performance of the sensor is investigated by injecting different concentrations (100 pM to 10 μM) of biotin prepared in PBS into the sensor microchannel and recording the corresponding resonant spectral and angular shifts in the reflectance spectra. Note that we used a single injection procedure and reflectance spectra of the sensor with different concentrations of biotin are recorded after a reaction time of 40 min. Also note that PBS is introduced into the microchannel to remove the unbound and weakly-attached biotin molecules before each injection of a new concentration of biotin. In Fig. 8.9a, b, we show the responses of the device during the detection of different concentrations of biotin in spectral and angular scan, respectively. As can be seen, a red shift in spectral scan and an increase in angle shift in angular scan is obtained with increasing biotin concentrations. It indicates the increase of refractive index change with the capture of biotin molecules. Indeed, a nonlinear variation of the shift with increase in biotin concentration is observed in both scans. Figure 8.9c represents the real-time binding kinetics of 10 pM biotin, where we observed a red shift and discrete steps in the resonance wavelength over time, which is due to the 0.2 nm discreteness in the wavelength sensitivity. The observed variability in the step size is due to statistical fluctuations in which larger or smaller numbers of binding events can occur. It has been reported that *type I* HMM-based sensor can detect biotin concentration as low as 10 μM using the same streptavidin-biotin affinity model [19]. Notably, GC-HMM based sensor demonstrates the detection of 10 pM biotin in PBS, which shows that sensitivity is increased further by six orders of magnitude by using *type II* HMM.

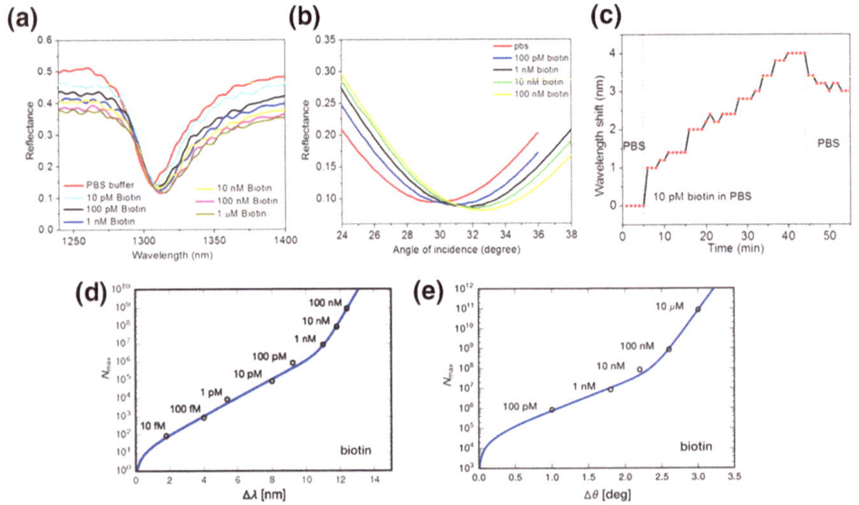

Fig. 8.9 Response of the HMM-based biosensors to the reaction of biotin–streptavidin binding in spectral and angular interrogation schemes: **a** Reflectance spectra of the sensor at 30° angle of incidence for different concentrations of biotin in PBS. **b** Reflectance spectra of the sensor device at 1250 nm for different concentrations of biotin in PBS. **c** The variation of wavelength shift in the presence of 10 pM biotin in PBS over time. Experimental results for the maximum number of biotin molecules, $N_{\max}$, adsorbed in the illuminated sensor areas in the long-time limit for the mode located at 1280 nm, **d** versus the corresponding wavelength shift $\Delta\lambda(c)$ and **e** versus the corresponding angular shift $\Delta\theta(c)$. Reproduced with permission from Ref. [22]

Then, we analyzed the sensitivity of the wavelength and angular shift to the number of adsorbed molecules on the sensor surface. For different concentrations c of biotin in PBS, we consider the saturation values of the wavelength ($\Delta\lambda$) and angular ($\Delta\theta$) shift. In particular, the shift of the resonance wavelength and angle depends on the number of bound molecules $N(c)$ in the sensing area. However, it is not directly possible to measure the precise value of $N(c)$, we can reliably estimate an upper bound $N_{\max}(c)$ based on the sensor parameters, such that the actual $N(c) \leq N_{\max}(c)$ at any concentration. By taking into account the microfluidic channel dimension and related sensor parameters, the calculated $N_{\max}(c) = 8.4\, c \times 10^{15}\ \mathrm{M}^{-1}$. As shown in Fig. 8.9d, e, there is a nonlinear relationship between experimental $N_{\max}(c)$ and $\Delta\lambda(c)$, and $N_{\max}(c)$ and $\Delta\theta(c)$, the shift in resonance wavelength and angle at concentration c. By using a phenomenological double-exponential fitting function shown in Eqs. (8.3) and (8.4), this behavior can be accurately reproduced. The blue curve in Fig. 8.9d represents the fitted spectral scan data using Eq. (8.3) and corresponding curve in Fig. 8.9e represents the fitted angular scan data using Eq. (8.4).

$$N_{\max} = A_1\left(e^{\Delta\lambda/\beta_1} - 1\right) + A_2\left(e^{\Delta\lambda/\beta_2} - 1\right) \tag{8.3}$$

$$N_{\max} = A_1\left(e^{\Delta\theta/\beta_1} - 1\right) + A_2\left(e^{\Delta\theta/\beta_2} - 1\right) \tag{8.4}$$

In Eqs. (8.3) and (8.4), A_1, A_2, β_1, β_2, are the fitting parameters. This double-exponential fitting function is chosen such that the limiting behavior at small $\Delta\lambda$ and $\Delta\theta$ is linear, so that one can expect in the case of very few adsorbed particles, which happens at lowest analyte concentration.

To conclude, the experimental realization of real-time label-free biosensing of small molecules at lower concentrations using the proposed GC-HMM platforms can be used to develop non-invasive liquid biopsies for POC clinical evaluation, early cancer screening and real-time diagnosis of diseases. In particular, exosomes can be regarded as novel disease markers for non-invasive detection and unravelling the process of exosome release could possibly yield new targets in anti-metastatic therapy [23]. Although their physiological roles are still under investigation, there is a growing need for reliable methods that can accurately isolate and detect exosomes from biological fluids. Over the years, a plethora of technologies and methods have been developed for the isolation of exosomes from biological fluids, including ultra-centrifugation, electron microscopy, fluorescence activated cell sorter (FACS), and conventional isolation kits based on buoyant density or polymer-based sedimentation [24]. Despite these advances in achieving enhanced sensitivity, no current technique has been integrated with standard bio-analytical systems for simultaneous profiling and quantification of the exosomes. We envision that exosomes can be detected even from bodily fluids and the detection limit can be improved by using the proposed plasmonic platform.

8.3.2 Spontaneous Emission Enhancement

The closed cavities such as microcavities and open cavities such as photonic crystals have widely been used as optical material geometries to better understand the cavity quantum electrodynamics and thus to study the spontaneous emission enhancement by harnessing the Purcell effect [25–28]. However, the optical systems based on microcavities and photonic crystal do not show high-quality factor resonances, which makes them inviable for broadband Purcell enhancement. Therefore, the design and development of novel optical systems are important to improve the broadband Purcell enhancement for various applications. Recently, HMMs have received much attention in this direction. The modifications in the spontaneous emission rates of organic dyes and quantum dots located in the close vicinity of the surface of HMMs have widely been investigated by many research groups [29–31]. However, a critical issue with HMM is that the Purcell enhancement recorded at far-field is limited by the non-radiative behavior of BPP modes in HMMs. Therefore, nanopatterned HMMs have been proposed to outcouple the non-radiative BPP modes of HMM to enhance the spontaneous emission rates of fluorescent molecules at far-field [32]. In this section, we discuss another approach to improve the spontaneous emission rate of quantum emitters by placing the emitter inside GC-HMM configuration, so that the grating coupler has the ability to outcouple and extract the non-radiative BPP modes at far-field [33]. In particular, we control the spontaneous emission enhancement by studying the interplay between the Purcell effect and the plasmon-exciton coupling in GC-HMMs doped with quantum emitters.

To investigate the fluorescence lifetime and decay rate of excitonic molecules coupled with the BPP modes of the GC-HMMs, time-resolved photoluminescence measurements are conducted. For this purpose, we developed an Ag/Al_2O_3 HMM that shows a *type II* hyperbolic dispersion above 430 nm wavelength (see Fig. 8.10a, b). We then fabricated a 2D Ag sub-wavelength diffraction grating on the Al_2O_3 space layer of HMM. If the emitter is directly placed on the HMM, it is not possible to couple the incident radiation to the emitter because of the impedance mismatch between air and HMM. However, in the GC-HMM configuration, the grating coupling outcouples the highly confined BPP modes from the structure to the far-field. In order to study the decay rate enhancement, a dye doped PMMA layer is deposited between 2D Ag diffraction grating and HMM, which is shown in Fig. 8.10a.

An ultrafast time resolved spectroscopy (TCSPC) setup with a time resolution of $\tau_{res} \leq 5$ ps is used to examine the HMM guided mode enhanced fluorescence lifetime of the dye molecules recorded at the far-field. Initially, the peak emission wavelength of Coumarin 500 dye dissolved PMMA layer is recorded at 470 nm with 380 nm excitation wavelength [33]. In the time-resolved experiments, the dye doped GC-HMM samples are excited using a pulsed laser with a wavelength of 380 nm, a pulse width of about 120 fs and a repetition rate of 4 MHz. To confirm the topological transition from elliptical to hyperbolic dispersion, the emission wavelength is varied. The measured lifetime data have been fitted using three exponential functions, $R(t) = B_1 e^{-t/\tau_1} + B_2 e^{-t/\tau_2} + B_3 e^{-t/\tau_3}$ with τ_i being the decay times. In the fitting

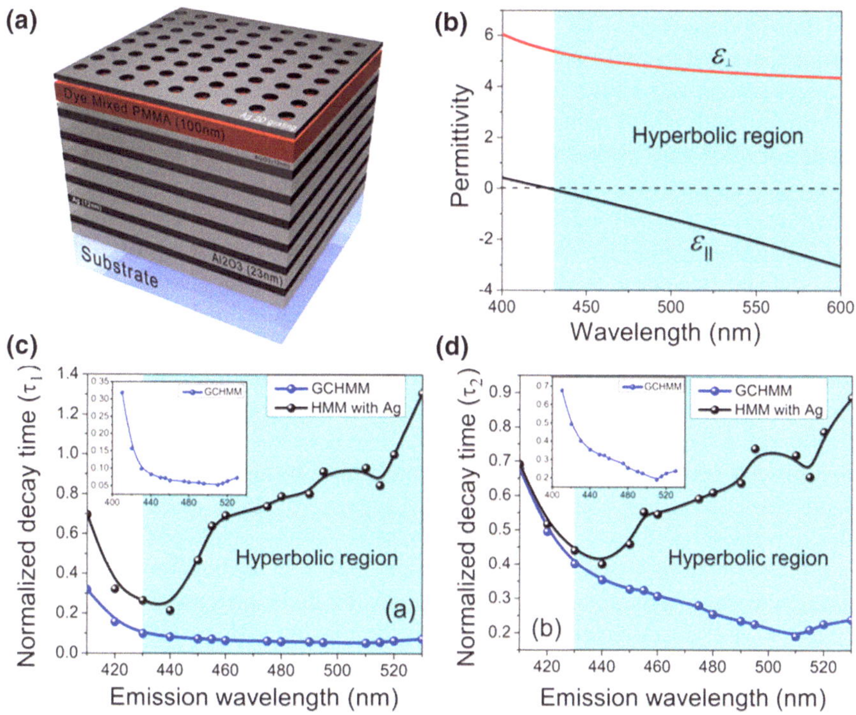

Fig. 8.10 **a** Schematic of fabricated Ag/Al_2O_3 GC-HMM, which consists of 6 pairs of Ag/Al_2O_3. **b** Real parts of effective permittivity of Ag/Al_2O_3 HMM determined with effective medium theory. The Ag/Al_2O_3 HMM shows hyperbolic dispersion at $\lambda \geq 450$ nm. **c**, **d** Lifetimes of dye on GC-HMM and Ag film deposited HMM normalized with respect to HMM sample, as a function of emission wavelength from elliptical to hyperbolic dispersion. Reproduced with permission from Ref. [33]

model, longer time, τ_3 is attributed to uncoupled dye molecules, located above the coupling distance from the HMM. Therefore, we consider shorter decay times, τ_1 and τ_2 to predict the decay rate enhancement because shorter decay times are related to molecules that are strongly coupled with HMM guided modes. In Fig. 8.10c, d, we show the measured fluorescent lifetimes (τ_1 and τ_2) of the short-living excitonic states of the emitters placed in the vicinity of GC-HMM as a function of emission wavelength. It is clear that the lifetime decreases in the hyperbolic region with increasing emission wavelength, which represents a clear signature of the topological transition from elliptical to hyperbolic dispersion.

The inverse of lifetime represents the decay rate enhancement. To further evidence that GC-HMM provides large decay rate enhancement compared to bare HMM, here we compare the decay rate enhancement of GC-HMM with that of an Ag film deposited HMM sample. Figure 8.10c, d respectively show the first (τ_1) and second (τ_2) decay time as a function of emission wavelength of both samples, which are normalized with respect to the corresponding lifetimes of bare HMM. In contrast to

Ag film deposited HMM, GC-HMM shows large decay rate enhancement especially throughout the hyperbolic spectral band of HMM. The Ag diffraction grating plays a major role in GC-HMM for enhancing the decay rate through strong coupling of emitters with HMM guided modes. In particular, both the *high-k* BPP and SPP modes of HMM are excited by the sub-wavelength Ag grating, as a result there is a strong overlap between these modes with the quantum emitters inside GC-HMM, which leads to a broadband enhancement of photonic density of states. In addition, Ag diffraction grating has the capability to scatter the BPP modes of HMM into well-defined free-space modes. Another reason for the decay rate enhancement is the sub-wavelength confinement of the emitter inside GC-HMM.

It is important to note that same decay rate variation is obtained for both decay times (τ_1 and τ_2). The preferential emission of *high-k* modes from GC-HMM is very clear as the lifetime values are higher in elliptical region and smaller in hyperbolic region. However, an Ag film deposited HMM does not supports the preferential emission of *high-k* modes as the lifetimes are randomly varying throughout the spectral band. The outcoupling ability of GC-HMM in terms of decay rate enhancement is shown in the inset of Fig. 8.10c, d. As can be seen, lifetime values of GC-HMM vary non-monotonically with emission wavelength; that is, in the hyperbolic region lifetime decreases to a minimum value at 510 nm, after that it increases with increase in emission wavelength. Since GC-HMM provides maximum decay rate at the outcoupling wavelength of 510 nm, we estimate the decay rate enhancement of GC-HMM at 510 nm. According τ_1, the experimentally estimated decay rate enhancement of GC-HMM in comparison to HMM and Ag film HMM is 18-fold and 17-fold, respectively. The radiative recombination rate is mainly responsible for the observed large spontaneous emission rate enhancement of GC-HMM, however, the non-radiative recombination rate is negligible because there is no direct contact between the fluorescent emitter and the metal layer. A most important advantage of sub-wavelength diffraction grating in GC-HMM is the ability to outcouple the BPP modes from HMM to far-field. In fact, the BPP field is highly confined in the HMM and it cannot be detected at far-field due to the evanescent nature of the confined field at the top surface of HMM. However, the diffraction grating converts this confined evanescent field into a propagating field for far-field detection. The influence of sub-wavelength diffraction grating on the spontaneous emission rate enhancement in GC-HMM is evident from the time-resolved photoluminescence measurements.

To provide further evidence to the enhanced emission rates observed in the experiments, decay rate enhancement of GC-HMM is numerically simulated. A semi-classical approach is used to study the decay rate enhancement analytically where the quantum emitters are considered as radiating point dipoles [34]. In this model, we consider a point dipole with dipole moment 'μ', is placed at a distance 'd' from an HMM and the decay rate enhancement with respect to vacuum is given by [34],

$$\beta = \frac{\Gamma}{\Gamma_0} = (1-\eta) + \frac{3\eta}{2k_0^3}$$
$$\text{Re}\int_0^\infty \exp(2ik_z d)\frac{k_{//}dk}{k_z|\mu^2|}\left\{\frac{1}{2}\mu_{//}^2[(1+r_s)k_0^2 - (1-r_p)k_z^2] + \mu_\perp^2(1+r_p)k_{//}^2\right\} \tag{8.5}$$

In Eq. (8.5), η is the intrinsic quantum yield of the emitter, k_0 is the vacuum wavevector of the medium in which emitter resides, $k_{//}$ is the wavevector parallel to the interface of the HMM, $k_z = \sqrt{k_0^2 - k_{//}^2}$ and r_s, r_p are the Fresnel reflection coefficients for *s*- and *p*-polarization, respectively, which are calculated with transfer matrix method. The frequency dependent complex permittivity of Ag is obtained from Drude model, $\varepsilon_{Ag}(\omega) = 1 - \left(\omega_p^2/(\omega^2 + i\nu\omega)\right)$ with $\omega_p = 11.5\ fs^{-1}$ and $\nu = 0.083\ fs^{-1}$ [34]. Since emitter resides in PMMA, the refractive index of PMMA is set to be 1.49 and the intrinsic quantum yield of Coumarin dye is assumed to be 0.98 [33]. The separation distance, *d* is set to be 12 nm.

In Fig. 8.11a, we show the numerically simulated decay rate enhancement (β) of HMM and GC-HMM. It is evident from the inset of Fig. 8.11a that HMM shows broadband decay rate enhancement in the hyperbolic spectral band ($\lambda \geq 430$ nm). The observed increase of decay rate enhancement in the hyperbolic region is due to the strong coupling between the emitter and the *high-k* modes of HMM. Note that strong coupling strength is obtained for shorter interaction distances (*d/λ*). As expected, GC-HMM shows improved decay rate enhancement throughout the hyperbolic band as compared to HMM. More importantly, GC-HMM exhibits large decay rate enhancement at a specific wavelength, which represents the outcoupling wavelength of GC-HMM with grating period 500 nm. The maximum β is obtained at 497 nm wavelength, which is 12-fold compared to that of an HMM. It implies that the simulation results correlate well with experimental results. However, there is a slight difference in emission wavelength at which maximum decay rate is obtained, which is due to the non-uniform periodicity of fabricated diffraction grating as compared to uniform periodicity considered in the simulation. In the numerical model, we introduced the periodicity of the diffraction grating in the grating coupling equation, $k_x = n_0 k_0 \sin\theta \pm m(2\pi/\Lambda_x) \pm n(2\pi/\Lambda_y)$ with $\Lambda_x = \Lambda_y = \Lambda$ for a diffraction grating with square lattice symmetry. Thus, the periodicity of the diffraction grating determines the maximum value of β.

The decay rate enhancement at the outcoupling wavelength can be tuned by changing the period of the diffraction grating. Therefore, in Fig. 8.11b we show the decay rate enhancement of GC-HMM with different grating periods. It is evident that the spectral position of the maximum β and broadband enhancement of β can be varied by changing the grating period. In particular, maximum decay rate enhancement with narrow band emission is observed for smaller period at $\Lambda = 475$ nm. This is because the efficient momentum matching is possible with smaller periods and thus,

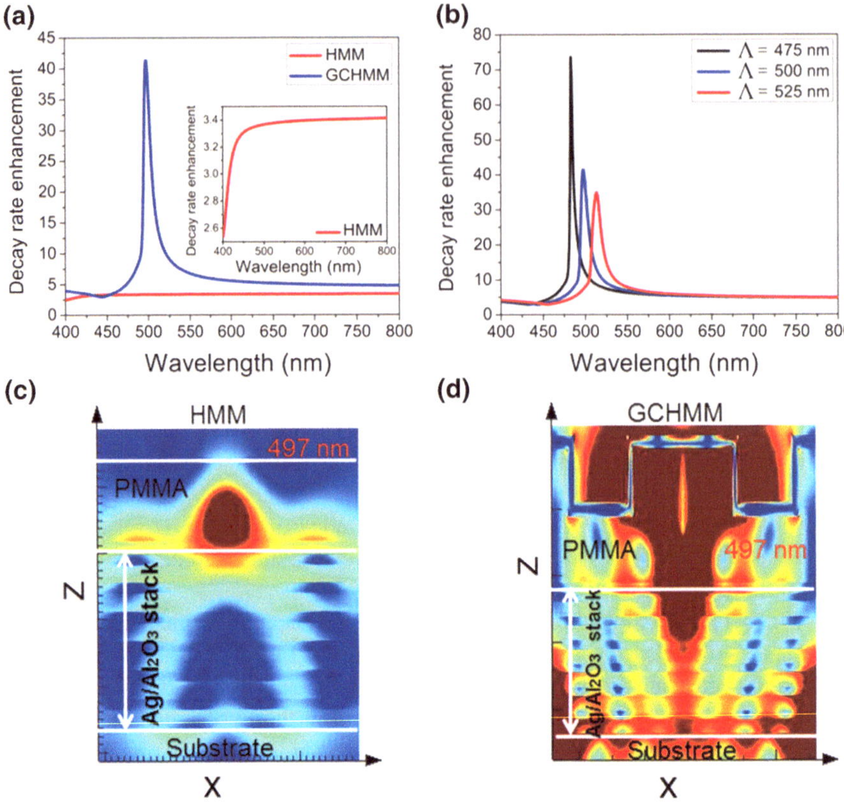

Fig. 8.11 **a** Numerically simulated decay rate enhancement of GC-HMM with period 500 nm and HMM. Enlarged plot of decay rate enhancement of HMM is shown in the inset of **a**. **b** Tuning of maximum decay rate enhancement of GC-HMM with grating period. Cross-sectional map of intensity distribution along the GC-HMM at 497 nm for **c** HMM and **d** GC-HMM. Reproduced with permission from Ref. [33]

there is a possibility of the outcoupling of most of the HMM guided modes. It is important to note that a single period diffraction grating can only outcouples a single wavelength even though HMM provides broadband decay rate enhancement in the hyperbolic region, which is observed in the simulations. However, the spectrum selective outcoupling of HMM *high-k* modes are possible by designing a proper chirped grating [35]. It indicates that the broadband outcoupling of maximum decay rate enhancement can be achieved by engineering the grating coupler geometry.

To further show the important role of grating coupling in GC-HMM for decay rate enhancement purpose, we performed finite difference time domain (FDTD) numerical simulations. In the simulation model, we place a point dipole above the metal-dielectric stack that represents the quantum emitter, with a specified emission wavelength resides in PMMA layer. In order to compare the results, we simulate the intensity field distribution along HMM and GC-HMM with a period of 500 nm.

Figure 8.11c, d, respectively show the cross-sectional intensity field distribution map of HMM and GC-HMM at the outcoupling wavelength of 497 nm. The enhanced intensity distribution on the substrate as well as above the grating (in air) is clearly visible in the case of GC-HMM, which indicates the outcoupling ability of GC-HMM. The observed intensity distribution supports the results shown in Fig. 8.11a that a maximum decay rate enhancement is obtained at 497 nm for 500 nm grating period. Moreover, it is important to notice that the enhanced intensity is obtained at the dye/HMM interface, so that higher plasmonic density of states can be outcoupled, as a result there is a strong overlap between the excited *high-k* modes and quantum emitter. Therefore, GC-HMM can provide a broadband enhancement of photonic density of states in addition to an outcoupling effect from GC-HMM. We further proposed a grating-coupled dye-embedded hyperbolic metamaterial (GC-DEHMM) geometry for improving the spontaneous emission enhancement of quantum emitters [36]. In comparison to a reference sample, about a 35-fold spontaneous emission decay rate enhancement of dye molecules is obtained using a 2D silver diffraction grating coupled with DEHMM.

One of the potential applications of GC-HMM is the development of an efficient, room temperature and polarized single photon source (SPS). Single photon source is one of the essential hardware components of quantum technologies, which has the capability to produce photons with antibunching features [25]. In past years, different types of room temperature SPS including single dye molecule, quantum dots, color centers in diamond, and spontaneous parametric down-conversion (SPDC) have been demonstrated [37–40]. However, these techniques are not suitable for practical applications. The improvement of emission rate enhancement of quantum emitter is another important issue to realize an efficient SPS. In addition, the doping of single photon emitters into photonic microcavities is a critical challenge to increase the SPS efficiency, polarization and bit-rate for room temperature operation [26]. The key features required for the practical realization of efficient single photon sources are: (i) large spontaneous emission rate, (ii) high radiative quantum efficiency with directional emission, and (iii) room-temperature and broadband operation with a specific polarization. By considering the results of GC-HMM, we envision that these features can be simultaneously achieved by coupling the *high-k* states in the HMM to photons propagating in air by special surface structuring.

8.3.3 *Multiband, Broad- and Narrow-Band Perfect Absorption and Absorption-Based Plasmonic Sensors*

It has been predicted that HMMs show blackbody behavior for both broadband and narrowband spectra [41–44]. Guclu et al. [41] have theoretically demonstrated that HMMs behave as electromagnetic absorbers for scattered fields. Later, enhanced absorption based on this concept was experimentally demonstrated by placing scatterers on top of the HMM, however, it did not show neither perfect nor narrow-band absorption [42]. The strong absorption of light in HMM is realized via the indefinite dispersion, even though HMM is made of low-loss dielectric materials and highly reflective noble metals. The divergence of the photonic density of states occurs due to this unusual hyperbolic dispersion, which results to a dramatic increase in the absorption of incident photons and it gets absorbed into the propagating modes of the hyperbolic medium. Specifically, *type II* HMMs are able to confine the electromagnetic field in a very small volume while allowing it to propagate at the same time in the form of BPPs since it shows overall positive permittivity along the bulk.

In this section, by exciting the BPPs of HMMs, we demonstrate that GC-HMMs open new venues for engineering electromagnetic absorption. In particular, we show polarization independent, wide angle range, multiband, broad-band, and narrow-band perfect absorption for a broad spectral range from visible to infrared using GC-HMMs. Moreover, we demonstrate a high-sensitivity absorption-based plasmonic sensor with an unprecedented high figure of merit using the proposed ultra-narrow band perfect absorber.

We designed and fabricated two different GC-HMMs such as a palladium (Pd) grating coupled Ag–TiO_2 HMM for broad-band perfect absorbers and an Au grating coupled Au–Al_2O_3 HMM for narrow-band perfect absorbers [45]. The fabricated Ag–TiO_2 HMM made of 12 alternating thin films of Ag and TiO_2 with same thickness (20 nm) of both Ag and TiO_2, whereas Au–Al_2O_3 HMM made of 16 alternating thin films of Au and Al_2O_3 with thickness of Au and Al_2O_3 is 15 nm and 30 nm, respectively. According to calculated uniaxial permittivity components based on EMT, Au–Al_2O_3 HMM shows *type II* hyperbolic dispersion in the optical frequencies, $\lambda > 520$ nm. However, Ag–TiO_2 HMM shows both *type I* ($325 \text{ nm} < \lambda < 410$ nm) and *type II* ($\lambda > 410$ nm) hyperbolic dispersion (for more details see ref. [45]). Interestingly, a strong discontinuity in $\varepsilon_\perp$, passing from a high negative value to a high positive value, and $\varepsilon_{//} = 0$ is realized at the transition wavelength (410 nm) of Ag–TiO_2 HMM. In order to integrate a 2D metallic diffraction grating with HMM to excite the BPP modes for broad and narrow-band perfect absorption, we fabricated a Pd sub-wavelength diffraction grating on the 10 nm thick TiO_2 spacer layer of the Ag–TiO_2 HMM and an Au sub-wavelength diffraction grating on the 10 nm thick Al_2O_3 spacer layer of the Au–Al_2O_3 HMM, respectively. The fabricated 2D sub-wavelength Pd (8 nm thick) and Au (20 nm thick) gratings have an average period of 500 nm and an average hole diameter of 200 nm.

By using a spectroscopic ellipsometer, we measured angular reflectance (R) and transmission (T). The absorption (A) is obtained from its relation to transmission and reflection, which is given by $A = 1 - T - R$, here we assume that the HMM surface is flat and the excited BPPs die out before re-scattering. As obtained angular absorption spectra of Pd GC-HMM and Au GC-HMM using TM polarization of light are shown in Fig. 8.12a, e, respectively. As can be seen, Pd GC-HMM shows broadband modes whereas Au GC-HMM shows narrowband modes. It indicates that a drastic change in the absorption properties occur if different metals are used for diffraction grating. We confirmed that the width of the excited modes depends on the contrast between the grating effective permittivity and the bulk effective parallel permittivity of HMM, that is, mode width decreases with decreasing contrast in permittivity. More specifically, the grating still exists while the HMM starts to vanish if the effective average permittivity of the grating approaches the effective average permittivity of the HMM [46]. The large contrast in effective permittivity between Pd grating and Ag–TiO_2 HMM results broadband absorption spectra of Pd GC-HMM, however, the narrowband absorption spectra of Au GC-HMM are due to the lower contrast in permittivity between Au grating and Au–Al_2O_3 HMM. In addition, the observed increase in mode bandwidth with increasing wavelength is due to the increase in permittivity contrast between the grating and the HMM as the wavelength increases [45].

Now we discuss the absorption properties of Pd GC-HMMs in detail. As shown in Fig. 8.12a, all the excited modes of Pd GC-HMM exhibit broad-band absorption above 98% for a wide range of angles of incidence (30°–60°). In fact, the perfect absorption (~100%) is realized at a particular angle of incidence for different excited modes. We observed that the perfect absorption angle decreased with decreasing excited mode resonance wavelength, as the mode at 2000, 950, 600 and 450 nm provides perfect absorption at 60°, 55°, 40°, and 35°, respectively. The observed perfect absorption in GC-HMM is due to the indefinite hyperbolic *k*-space. As mentioned above, it dramatically increases the photonic density of states, which leads to enhance the absorption of incident light inside the HMM especially in the *type II* region, in the form of BPPs. In particular, the density of states of HMMs are indefinite in the hyperbolic dispersion frequency range. Therefore, GC-HMM offers flexibility in engineering absorption by exciting the BPPs depending on the predesigned grating period of sub-wavelength gratings. To further prove our claims, we performed intensity field distribution simulations using FDTD. In Fig. 8.12b, we show intensity distribution along the Pd GC-HMM at excited BPP mode resonances. The intensity field distribution of a control sample (Pd grating on 1 pair of Ag–TiO_2) for the same resonance wavelengths are shown in Fig. 8.12c. One can see that GC-HMM shows strong absorption of light in the HMM for all the resonance wavelengths of BPP modes due to hyperbolic dispersion. The field is tightly confined inside the multilayer of Pd GC-HMM, whereas the field is mostly concentrated around the grating of a control sample.

Realizing polarization independent absorption using metamaterial perfect absorbers (MPAs) is a challenging task. However, fabricating a combination of metamaterial absorbers rotated by 90° is required to overcome this problem. Nevertheless, a key feature of the designed GC-HMM is that it exhibits polarization independent absorption. Since we use 2D sub-wavelength grating geometry, GC-HMM inherently exhibits a $\pi/2$ rotational symmetry. Note that MPAs work based on the impedance matching conditions, absorption of the TE mode decreases at higher angles of incidence, whereas the absorption of the TM mode is left unaffected. The reason is that the magnetic field creates parallel and antiparallel currents, which generates the required left-handed magnetic response of impedance matching MPAs. In particular, the magnetic field lies in the same plane as the propagation *k* vector, and the component along the plane decreases at higher angles for TE modes [47]. The wide angle polarization independent absorption properties of Pd GC-HMM is shown in

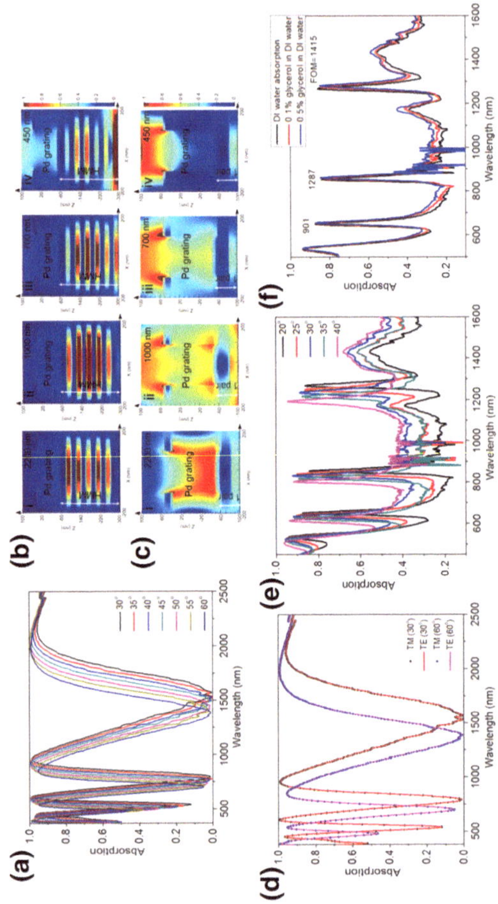

Fig. 8.12 **a** Angular absorption spectra of Pd GC-Ag/TiO_2 HMM for TM polarization. Cross-sectional intensity field distribution maps of **b** Pd GC-Ag/TiO_2 HMM at BPP mode resonance wavelengths, and **c** control sample (Pd GC-1 pair of Ag/TiO_2) at same wavelengths. **d** Absorption spectra of Pd GC-Ag/TiO_2 HMM at 30° and 60° angle of incidence for TM and TE polarization. **e** Angular absorption spectra of Au GC-Au/Al_2O_3 HMM for TM polarization. **f** Refractive index sensing of proposed plasmonic absorption sensor using different weight percentages of glycerol in DI water. Reproduced with permission from Ref. [45]

Fig. 8.12d, which clearly shows that such effect does not occur in GC-HMM since the mode profile and absorption intensity for TE and TM polarization are exactly same for both the angles.

Both broadband and ultra-narrow band absorbers are equally important for many potential applications, however, little efforts have been made to realize ultra-narrow band absorbers. This is because it is a challenging task to realize ultra-narrow band plasmonic absorbers in the optical spectral band due to the high optical losses associated with noble metals. To overcome this issue, we use Au GC-HMMs to obtain ultra-narrow band absorption for multiple bands from visible to near infrared frequencies. Here, multi narrow-band modes are realized by coupling 20 nm Au grating with an Au–Al_2O_3 HMM, where the contrast in permittivity between Au grating and Au–Al_2O_3 HMM is much smaller. The lower grating modulation amplitude of Au grating also contributes to achieve narrow band modes. In Fig. 8.12e we show the TM polarized absorption spectra of Au GC-HMM at different angles of incidence, where the excitation of BPP modes results four ultra-narrow band absorption peaks. The measured maximum absorption and FWHM at 40° angle of incidence for the mode at 1192, 805 and 614 nm are 95% and 23 nm, 92% and 18 nm, and 94% and 17 nm, respectively. In fact, Au GC-HMM provides ultra-narrow band super absorption (above 90% absorption).

We further demonstrate an application of Au GC-HMM as a plasmonic absorber sensor by exploiting the ultra-narrow band absorption properties of Au GC-HMM absorbers [45]. In this case, a small change in the refractive index of environment results in measurable changes in the optical properties of the sensor. Specifically, the absorption features of the Au GC-HMM can change, which indicates a change in the polarizability of the superstrate. For an absorber plasmonic sensor [47], FOM = $\max\left[\frac{dI(\lambda)/dn}{I(\lambda)}\right]$ with $dI(\lambda)/I(\lambda)$ is the relative intensity change at a fixed wavelength induced by a refractive index change dn and $I(\lambda)$ corresponds to the intensity where FOM is maximized such that a relative intensity change is caused by a change in refractive index [45]. The previously reported maximum FOM for a plasmonic absorber sensor is 87 Ref. [47]. To demonstrate the sensing properties of the proposed absorber sensor, we integrated a microfluidic channel with our Au GC-HMMs. We then injected different concentrations of glycerol dissolved in distilled (DI) water as the known refractive index analyte and monitored the absorption changes with change in refractive index of analyte. In Fig. 8.12f, we show the shift in the absorption peak for 0.1 and 0.5% of glycerol concentrations with respect to DI water. The wavelength shift and corresponding refractive index sensitivity recorded for 0.5% of glycerol concentration with respect to DI water are 14 nm and 23,333 nm/RIU, 8 nm and 13,333 nm/RIU, and 6 nm and 10,000 nm/RIU for the mode at 1290, 870 and 660 nm, respectively. Moreover, the obtained FOMs for the proposed plasmonic absorber sensor are 1415, 1287, and 901 for the mode at 1290, 870 and 660 nm, respectively. The maximum FOM obtained for the mode at 1290 nm is about 16-fold higher than that of a previously reported system [47].

It is important to notice that the maximum intensity change does not occur at the local maxima of the absorption modes, however it happens at a nearby wavelength where the slope of the mode is highest. The presented ultra-narrow band absorber sensor based on GC-HMMs represent remarkable features required for new generation biosensors. An important advantage of plasmonic absorber sensor is that the sensing measurement can be performed by using a single wavelength and a single detection angle, since the detection mechanism is based on strong intensity variation for a specific wavelength at a specific incidence angle. More importantly, the narrow-band absorption sensors that operate at optical frequencies could revolutionize the current biodetection approaches, as the binding events can be directly visualized due to the remarkable absorption changes of the sensor.

References

1. Newman WD, Cortes CL, Atkinson J, Pramanik S, DeCorby RG, Jacob Z (2015) Ferrell–Bereman modes in plasmonic epsilon-near-zero media. ACS Photonics 2:2–7
2. Xu C, Xu J, Song G, Zhu C, Yang Y, Agarwal GS (2016) Enhanced displacements in reflected beams at hyperbolic metamaterials. Opt Exp 24:21767–21776
3. Avrutsky I, Salakhutdinov I, Elser J, Podolskiy V (2007) Highly confined optical modes in nanoscale metal-dielectric multilayers. Phys Rev B 75:241402
4. Cortes CL, Newman W, Molesky S, Jacob Z (2012) Quantum nanophotonics using hyperbolic metamaterials. J Opt 14:063001
5. Zhukovsky SV, Kidwai O, Sipe JE (2013) Physical nature of volume plasmon polaritons in hyperbolic metamaterials. Opt Exp 21:14982
6. Ishil S, Kildishev AV, Narimanov E, Shalaev VM, Drachev VP (2013) Sub-wavelength interference pattern from volume plasmon polaritons in a hyperbolic medium. Laser Photonics Rev 7:265–271
7. Yan W, Shen L, Ran L, Kong JA (2007) Surface modes at the interfaces between isotropic media and indefinite media. J Opt Soc Am A 24:530–535
8. Sreekanth KV, De Luca A, Strangi G (2013) Experimental demonstration of surface and bulk plasmon polaritons in hypergratings. Sci Rep 3:3291
9. Sreekanth KV, De Luca A, Strangi G (2014) Excitation of volume plasmon polaritons in metal-dielectric metamaterials using 1D and 2D diffraction gratings. J Opt 16:105103
10. Sreekanth KV, Ouyang Q, Sreejith S, Zeng S, Lishu W, Ilker E, Dong W, ElKabbash M, Ting Y, Lim CT, Hinczewski M, Strangi G, Yong KT, Simpson RE, Singh R (2019) Phase change material-based low-loss visible frequency hyperbolic metamaterials for ultra-sensitive label-free biosensing. Adv Opt Mater 1900081
11. Zeng S, Sreekanth KV, Shang J, Yu T, Chen CK, Yin F, Baillargeat D, Coquet P, Ho HP, Kabashin AV, Yong KT (2015) Graphene-gold metasurface architectures for ultrasensitive plasmonic biosensing. Adv Mater 27:6163–6169
12. Yin X, Hesselink L (2006) Goos-Hänchen shift surface plasmon resonance sensor. Appl Phys Lett 89:261108
13. Wan Y, Zheng Z, Kong W, Zhao X, Liu Y, Bian Y, Liu J (2012) Nearly three orders of magnitude enhancement of Goos-Hanchen shift by exciting Bloch surface wave. Opt Exp 20:8998–9003
14. Brolo AG (2012) Plasmonics for future biosensors. Nat Photon 6:709–713
15. Homola J, Yee SS, Gauglitz G (1999) Surface plasmon resonance sensors: review. Sens Actuators B 54:3
16. Wu C, Adato R, Arju N, Yanikk AA, Altug H, Shvets G (2012) Fano-resonant asymmetric metamaterials for ultrasensitive spectroscopy and identification of molecular manolayers. Nat Mater 11:69–75
17. Zeng S, Baillargeat D, Hod HP, Yong KT (2014) Nanomaterials enhanced surface plasmon resonance for biological and chemical sensing applications. Chem Soc Rev 43:3426–3452

18. Kravets VG, Schedin F, Jalil R, Britnell L, Gorbachev RV, Ansell D, Thackray B, Novoselov KS, Geim AK, Kabashin AV, Grigorenko AN (2013) Singular phase nano-optics in plasmonic metamaterials for label-free single molecule detection. Nat Mater 12:304–309
19. Kabashin AV, Evans P, Pastkovsky S, Hendren W, Wurtz GA, Atkinson R, Pollard R, Podolskiy VA, Zayats AV (2009) Plasmonic nanorod metamaterials for biosensing. Nat Mater 8:867–871
20. Sreekanth KV, Alapan Y, ElKabbash M, Ilker E, Hinczewski M, Gurkan UA, De Luca A, Strangi G (2016) Extreme sensitivity biosensing platform based on hyperbolic metamaterials. Nat Mater 15:621–627
21. Sreekanth KV, ElKabbash M, Wen AM, Ilker E, Hinczewski M, Gurkan UA, Steinmetz NF, Strangi G (2016) Enhancing the angular sensitivity of plasmonic sensors using hyperbolic metamaterials. Adv Opt Mater 4:1767–1772
22. Strangi G, Sreekanth KV, Elkabbash M (2017) Hyperbolic metamaterial-based ultrasensitive plasmonic biosensors for early-stage cancer detection. In: Chandra P, Tan Y, Singh S (eds) Next generation point-of-care biomedical sensors technologies for cancer diagnosis. Springer, Singapore
23. Poudineh M, Sargent EH, Pantel K, Kelley SO (2018) Profiling circulating tumour cells and other biomarkers of invasive cancers. Nat Biomed Eng 2:72–84
24. Boriachek K, Islam MN, Möller A, Salomon C, Nguyen NT, Hossain MSA, Yamauchi Y, Shiddiky MJA (2018) Biological functions and current advances in isolation and detection strategies for exosomes nanovesicles. Small 14:1702153
25. Lounis B, Orrit M (2005) Single-photon sources. Rep Prog Phys 68:1129
26. Hughes S (2004) Enhanced single-photon emission from quantum dots in photonic crystal waveguides and nanocavities. Opt Lett 29:2659
27. Sreekanth KV, Biaglow T, Strangi G (2013) Directional spontaneous emission enhancement in hyperbolic metamaterials. J Appl Phys 114:134306
28. Purcell EM (1946) Spontaneous emission probabilities at radio frequencies. Phys Rev 69:681
29. Jacob Z, Smolyaninov II, Narimanov EE (2012) Broadband Purcell effect: radiative decay engineering with metamaterials. Appl Phys Lett 100:181105
30. Ni X, Ishii S, Thoreson MD, Shalaev VM, Han S, Lee S, Kildishev AV (2011) Loss-compensated and active hyperbolic metamaterials. Opt Exp 19:25242–25254
31. Krishnamoorthy HNS, Jacob Z, Narimanov E, Kretzschmar I, Menon VM (2012) Topological transitions in metamaterials. Science 336:205
32. Lu D, Kan JJ, Fullerton EE, Liu Z (2014) Enhancing spontaneous emission rates of molecules using nanopatterned multilayer hyperbolic metamaterials. Nat Nanotech 9:48–53
33. Sreekanth KV, Hari KK, De Luca A, Strangi G (2014) Large spontaneous emission rate enhancement in grating coupled hyperbolic metamaterials. Sci Rep 3:3291
34. Ford GW, Weber WH (1984) Electromagnetic interactions of molecules with metal surfaces. Phys Rep 105:227403
35. Bartoli FJ, Gan Q (2011) Surface dispersion engineering of planar plasmonic chirped grating for complete visible rainbow trapping. Appl Phys Lett 98:251103
36. Krishna KH, Sreekanth KV, Strangi G (2016) Dye-embedded and nanopatterned hyperbolic metamaterials for spontaneous emission rate enhancement. J Opt Soc Am B 33:1038–1043
37. Brokmann X, Giacobino E, Dahan M, Hermier JP (2004) Highly efficient triggered emission of single photons by colloidal CdSe/ZnS nanocrystals. Appl Phys Lett 85:712
38. Beveratos A, Kühn S, Brouri R, Gacoin T, Poizat J, Grangier P (2002) Room temperature stable single-photon source. Eur Phys J D 18:191
39. Soujaeff A, Takeuchi S, Sasaki K, Hasegawa T, Matsui M (2007) Heralded single photon source at 1550 nm from pulsed parametric down conversion. J Mod Opt 54:467
40. Treussart F, Alleaume R, Courty J, Roch J (2004) Emission properties of a single photon source. Phys Scr T 112:95
41. Guclu C, Campione S, Capolino F (2012) Hyperbolic metamaterial as super absorber for scattered fields generated at its surface. Phys Rev B 86:205130
42. Narimanov EE, Li H, Barnakov YA, Tumkur TU, Noginov MA (2013) Reduced reflection from roughened hyperbolic metamaterial. Opt Exp 21:14956

43. Tumkur TU, Kitur JK, Chu B, Gu L, Podolskiy VA, Narimanov EE, Noginov MA (2012) Control of reflectance and transmittance in scattering and curvilinear hyperbolic metamaterials. Appl Phys Lett 101:091105
44. Halterman K, Elson JM (2014) Near-perfect absorption in epsilon-near-zero structures with hyperbolic dispersion. Opt Exp 22:7337
45. Sreekanth KV, Alapan Y, Rashed AR, Gurkan UA, Strangi G (2016) A multiband perfect absorber based on hyperbolic metamaterials. Sci. Rep. 6:26272
46. Wang SS, Magnusson R (1993) Theory and applications of guided-mode resonance filters. Appl Opt 32:2606–2613
47. Liu N, Mesch M, Weiss T, Hentschel M, Giessen H (2010) Infrared perfect absorber and its application as plasmonic sensor. Nano Lett 10:2342–2348

Chapter 9
Graphene and Topological Insulator-Based Active THz Hyperbolic Metamaterials

In this chapter, we demonstrate negative group refraction properties of graphene- and topological insulator-based HMMs. Graphene-based HMMs show *type II* hyperbolic dispersion and topological insulator such as Bi_2Se_3 shows both *type I* and *type II* hyperbolic dispersion at terahertz (THz) frequencies. In *type II* spectral region, negative group index of refraction at higher angles of incidence is obtained for both HMMs. However, an all angle negative group index of refraction is observed in the *type I* hyperbolic dispersion of Bi_2Se_3. Moreover, we demonstrate broadband tunable hyperbolic dispersion from THz to mid infrared frequencies and a widening of the negative group index spectral band using topological insulator-dielectric insulator based superlattice structures.

9.1 Introduction

It was reported that different material combinations are required to realize hyperbolic metamaterials from ultraviolet (UV) to terahertz (THz) frequencies. For UV and visible frequencies, the hyperbolic dispersion is usually realized by combining noble metals, such as Au and Ag, and dielectrics such as SiO_2, Al_2O_3, MgF_2, TiO_2, Si and PMMA [1–4]. However, TiN has been proposed to replace the metallic layers in visible photonics applications since noble metal-based HMMs are severely limited due to the strong dispersive nature of metals at visible frequencies [5]. Heavily doped oxide semiconductors, such as Al:ZnO, InGaAs, and SiC have been used as an alternative plasmonic material to realize hyperbolic dispersion at infrared frequencies because noble metals are unsuitable for this spectral regime due to their high reflectivity [6, 7]. The low-loss and tunable plasmonic response of graphene suggest that graphene-dielectric stacks could be a suitable choice to realize hyperbolic dispersion and active HMMs in infrared and THz frequencies [8, 9]. It has also been reported that active HMMs can be realized using tunable materials such as electride materials [10], hexagonal boron nitride [11], and phase change oxides and chalcogenides [12, 13].

Sreekanth K. V. et al., *New Directions in Thin Film Nanophotonics*, Progress in Optical Science and Photonics 6, https://doi.org/10.1007/978-981-13-8891-0_9

9.2 Graphene-Based Hyperbolic Metamaterials

In recent years, graphene plasmonics has generated great interest among scientific community because of the ability of graphene to tune the plasmon dispersion by varying the chemical potential via external gating or doping [14]. Graphene is a single two-dimensional plane of carbon atoms forming a hexagonal lattice that supports transverse magnetic (TM) surface plasmons similar to the surface plasmons of metal/dielectric systems [15]. However, for optical and near-infrared frequencies, graphene is not a promising material to replace metals for constructing metamaterials due to its tremendous losses. Nonetheless, graphene is a potential choice to develop hyperbolic metamaterials in the mid-infrared and THz frequencies when it is combined with dielectric materials. Such graphene-based hyperbolic metamaterials may offer new venue for achieving high-speed dynamically tunable metamaterials, empowering the future applications of metamaterials in information processing and communications. Many research groups have investigated the optical properties of graphene and it is predicted that intraband absorption dominates at terahertz frequencies, which describes the dispersion of plasmons in graphene [16].

In Fig. 9.1a, the designed graphene-based HMM is shown, consisting of 6 bilayers of graphene/dielectric stack. The conductivity of graphene takes a Drude-like form in THz frequencies because of the dominance of intraband scattering in highly doped graphene [17]. Therefore, we model each single layer graphene (SLG) as a surface conducting sheet with conductivity, $\sigma_g = ie^2\mu/\left(\pi\hbar^2(\omega + i/\tau)\right)$ with ω is the angular frequency, e is the electron charge, τ is the electron relaxation time and μ is the chemical potential, which is a function of carrier density ($\mu \propto \sqrt{N_c}$). Here, we neglected interband absorption because the operating frequency is less than the chemical potential ($\omega < \mu$). The complex permittivity of graphene can be obtained from the expression, $\varepsilon_g = 1 + i(\sigma_g\eta_0/k_0t_g)$, where; $\eta_0 \approx 377$ is the impedance of air, t_g is the effective graphene thickness and $k_0 = \omega/c$ is the vacuum wavevector with c the speed of light [17]. In our calculations, the effective thickness of graphene was set to 0.5 nm and the electron relaxation time was assumed to be 1 ps, in order to account the scattering losses from the acoustic phonons [17]. More importantly, we varied the chemical potential of graphene to tune the complex permittivity.

Since the designed graphene-based HMM belongs to effective medium approximations, we used effective medium theory to study the wave propagation. The THz graphene-based HMM is an anisotropic medium with uniaxial dielectric tensor components, $\varepsilon_{xx} = \varepsilon_{yy} = \varepsilon_{//}$ and $\varepsilon_{zz} = \varepsilon_{\perp}$, which are approximated as,

$$\varepsilon_{//} = \frac{t_g\varepsilon_g + t_d\varepsilon_d}{t_g + t_d} \tag{9.1}$$

$$\varepsilon_{\perp} = \frac{\varepsilon_g\varepsilon_d(t_g + t_d)}{t_g\varepsilon_d + t_d\varepsilon_g} \tag{9.2}$$

where t_d and ε_d are the thickness and dielectric permittivity of dielectric material, which in our calculations are taken to be 50 nm and 2.25, respectively. In comparison

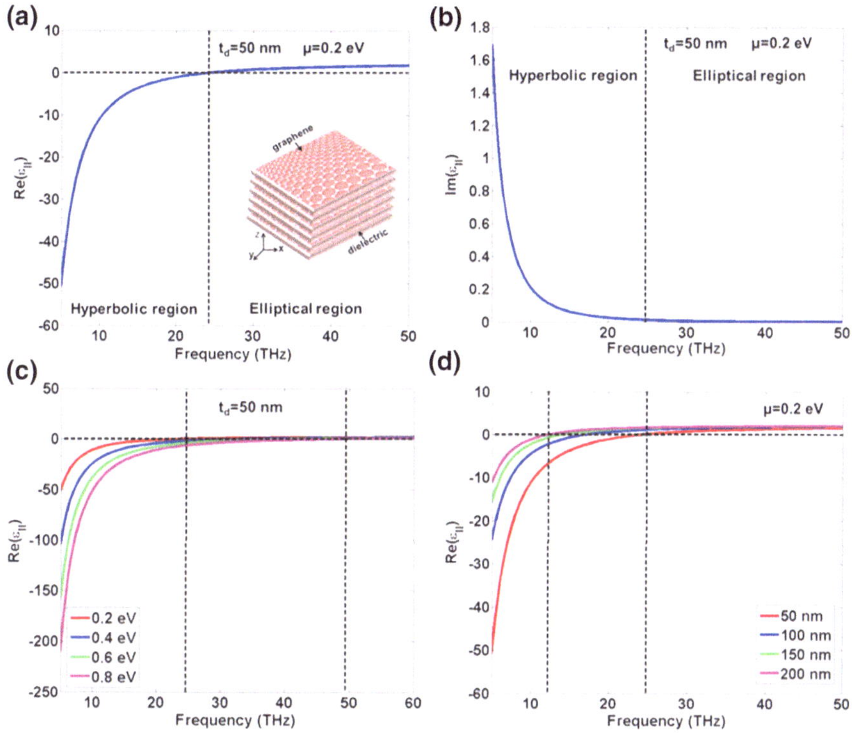

Fig. 9.1 **a** The frequency dependent variation of real parts of $\varepsilon_{//}$. Inset shows the schematic of graphene-based HMM consisting of 12 alternating layers of graphene and dielectric with graphene on top. **b** The frequency dependent variation of imaginary parts of $\varepsilon_{//}$. In **a** and **b**, $t_d = 50$ nm and $\mu = 0.2$ eV. The frequency dependent variation of real parts of $\varepsilon_{//}$ with **c** chemical potential, and **d** dielectric thickness. In **c**, $t_d = 50$ nm and in **d**, $\mu = 0.2$ eV. Reproduced from Sreekanth et al. [23], with the permission of AIP Publishing

to dielectric layer thickness, graphene thickness is negligible, therefore the dielectric tensor components in the perpendicular direction can be considered equal to the permittivity of dielectric layer ($\varepsilon_{\perp} = \varepsilon_d$). However, the dielectric tensor in the parallel direction (xy-plane) varies with THz frequencies.

The variation of real and imaginary parts of $\varepsilon_{//}$ with frequency is shown in Fig. 9.1a, b, respectively. As shown in Fig. 9.1a, the hyperbolic dispersion is obtained for certain frequencies in which values of the real parts of $\varepsilon_{//}$ are negative. In the elliptical region, values of the real parts of $\varepsilon_{//}$ result positive and approaching the permittivity of dielectric layer for higher frequencies. The frequency at which transition from elliptical to hyperbolic dispersion occurs at 24 THz for a chemical potential of $\mu = 0.2$ eV, which is considered to be the critical frequency. It is also clear from Fig. 9.1b that the values of the imaginary parts of $\varepsilon_{//}$ are decreasing in the hyperbolic region with increasing frequency. More importantly, these values are zero in the elliptical region. As a result, graphene-based HMM behaves like a dielectric medium in the elliptical region. An important advantage of graphene-based HMM is that the hyperbolic spectral band can be tuned by changing the chemical potential of graphene. In Fig. 9.1c, we illustrate the tuning of hyperbolic spectral band by increasing the chemical potential of graphene. In particular, the obtained critical frequencies are 24 and 49 THz for $\mu = 0.2$ eV and $\mu = 0.8$ eV, respectively. It shows that the broadband tuning of critical frequency is possible by slightly changing the chemical potential of graphene. In addition, the hyperbolic dispersion can be further tuned by changing the dielectric layer thickness. It is evident from Fig. 9.1d that the hyperbolic band shifts towards lower frequencies when the dielectric layer thickness increases, due to the increase of unit cell size of metamaterial. The observed critical frequencies are 24 and 12 THz for $t_d = 50$ nm and $t_d = 200$ nm, respectively.

9.3 Van der Waals Superlattice-Based Hyperbolic Metamaterials

The topological insulators (TIs) such as Bi_2Se_3 and Bi_2Te_3 are layered material, which are characterized by unusual gapless edge or surface states, and a full insulating gap in the bulk [18]. Very recently, TIs have received much attention for THz plasmonics [19] and investigated for active HMM applications. In particular, quintuple layered Bi_2Se_3 was theoretically investigated to realize hyperbolic dispersion at THz frequencies [20]. TI-based HMMs support hyperbolic phonon polaritons, which are highly directional and deeply subdiffractional modes. Furthermore, the dispersion relation of TI-based HMMs can be tuned by doping their surface states [20]. Therefore, high-speed dynamically tunable THz photonic devices could be realized using TI-based active HMMs.

Note that Bi_2Se_3 is composed of Se-Bi-Se-Bi-Se quintuple atomic planes that are separated by insulating Van der Waals bonds. The layered structure of Bi_2Se_3 is shown in the inset of Fig. 9.2b. The layered structure exhibits strong anisotropy of their phonon modes. It is known that the dominant (x-y)-axis and z-axis phonon frequencies of Bi_2Se_3 are 1.92 and 4.05 THz, respectively [21]. Bi_2Se_3 exhibits extremely anisotropic dielectric permittivity since these phonon mode frequencies are separated by a factor of two. For a certain range of THz frequency band, the real parts of the uniaxial permittivity components, $\varepsilon^x(\omega)$ and $\varepsilon^z(\omega)$ are indefinite. That is, $\varepsilon^x(\omega) \cdot \varepsilon^z(\omega) < 0$. As a result, for transverse magnetic (TM) polarization, Bi_2Se_3 exhibits hyperbolic dispersion at THz frequencies, $\frac{(k^x)^2+(k^y)^2}{\varepsilon^z(\omega)} + \frac{(k^z)^2}{\varepsilon^x(\omega)} = \frac{\omega^2}{c^2}$ and supports hyperbolic phonon polaritons.

We used the following model to determine the uniaxial permittivity components of Bi_2Se_3 [20],

$$\varepsilon^\alpha(\omega) = \varepsilon^\alpha_\infty + \sum_{j=1,2} \frac{\omega^{\alpha 2}_{p,j}}{\omega^{\alpha 2}_{to,j} - \omega^2 - i\gamma^\alpha_j \omega}, \quad \alpha = x, z \tag{9.3}$$

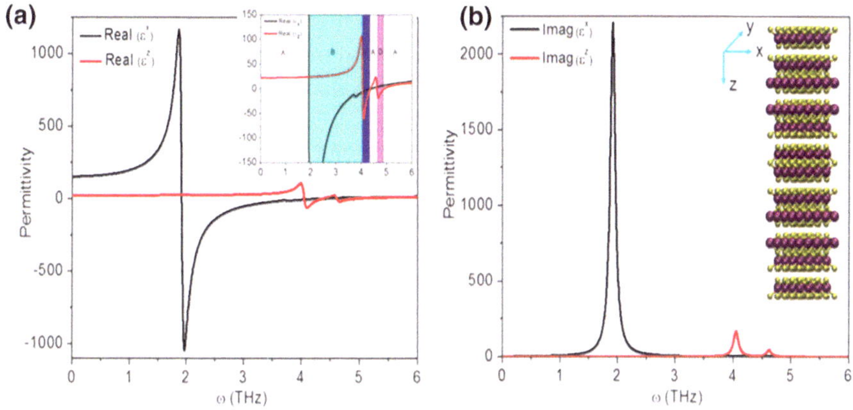

Fig. 9.2 **a** Real, **b** imaginary frequency dependent uniaxial permittivity components of Bi_2Se_3 ($\varepsilon^x(\omega)$ and $\varepsilon^z(\omega)$). The schematic of quintuple-layered Bi_2Se_3 is illustrated in the inset of (**b**). Reproduced with permission from Sreekanth and Simpson [24]

The value of components used in the calculations are, $\varepsilon^x_\infty = 29$, $\varepsilon^z_\infty = 17.4$, $\omega^x_{to,1} = 1.92$ THz, $\omega^x_{p,1} = 21.1$ THz, $\omega^x_{to,2} = 3.75$ THz, $\omega^x_{p,2} = 1.65$ THz, $\omega^z_{to,1} = 4.05$ THz, $\omega^z_{p,1} = 8.5$ THz, $\omega^z_{to,2} = 4.61$ THz, $\omega^z_{p,2} = 4.67$ THz and $\gamma^\alpha_j = 0.105$ THz.

The obtained real and imaginary parts of permittivity components ($\varepsilon^x(\omega)$ and $\varepsilon^z(\omega)$) are shown in Fig. 9.2a, b respectively. The real permittivity components alter sign from positive to negative with frequency, whilst both imaginary components are always positive. The enlarged plot in the inset of Fig. 9.2a shows, region A represents dielectric band ($\varepsilon^x(\omega) > 0$ and $\varepsilon^z(\omega) > 0$), region B represents *type II* hyperbolic band ($\varepsilon^x(\omega) < 0$ and $\varepsilon^z(\omega) > 0$), region C represents Reststrahlen band ($\varepsilon^x(\omega) < 0$ and $\varepsilon^z(\omega) < 0$) and region D represents *type I* hyperbolic band ($\varepsilon^x(\omega) > 0$ and $\varepsilon^z(\omega) < 0$). It implies that Bi_2Se_3 supports both *type I* and *type II* hyperbolic dispersion at THz frequencies, however above discussed graphene-based HMM shows only *type II* hyperbolic dispersion. The *type II* hyperbolic band of Bi_2Se_3 has a maximum bandwidth of 2.14 THz, which belongs to the bandwidth difference between two phonon frequencies at 1.91 and 4.05 THz. However, the Reststrahlen band (4.05–4.4 THz) and *type I* hyperbolic band (4.6–4.9 THz) are narrow with a bandwidth of 0.35 and 0.3 THz, respectively.

9.4 Negative Refraction in THz Hyperbolic Metamaterials

To achieve negative refraction in thin wire- and split ring-based metamaterials, double resonances, ε and μ are simultaneously less than zero, and extreme nanofabrication techniques are required. However, in HMMs, the optical anisotropy plays a major role to realize negative refraction, in addition to a single resonance causing ε to be less than zero. Heavily doped oxide semiconductors-based HMMs have been used to demonstrate negative refraction in mid-infrared frequencies [7].

It is known that in an anisotropic material the electric field vector, ***E*** and electric displacement vector, ***D*** are not usually parallel. As a result, the Poynting vector, ***S*** (direction of energy flow) and wavevector ***k*** (along the wavefront normal) are not parallel. Therefore, anisotropic materials exhibit negative refraction with respect to ***S***, but positive refraction with respect to ***k*** when the boundary condition in which tangential component of ***k*** is conserved at the interface [7]. In addition, the directions of ***k*** and ***S*** depend on the effective phase (n_p) and group (n_g) indices of refraction, respectively. For low material absorption, the effective phase and group index of an anisotropic medium can be obtained from Snell's law as [22],

$$n_p = \sqrt{\varepsilon_{//} + \left(1 - \frac{\varepsilon_{//}}{\varepsilon_{\perp}}\right)\varepsilon_0 \sin^2 \theta_{inc}} \tag{9.4}$$

$$n_g = \frac{\varepsilon_{\perp}}{\varepsilon_{//}}\sqrt{\varepsilon_{//} - \frac{\varepsilon_{//}}{\varepsilon_{\perp}}\left(1 - \frac{\varepsilon_{//}}{\varepsilon_{\perp}}\right)\varepsilon_0 \sin^2 \theta_{inc}} \tag{9.5}$$

where ε_0 is the permittivity of air. It is clear from the above equations that the effective phase and group index are related to angles of incidence (θ_{inc}).

9.4.1 Negative Refraction in Graphene-Based HMMs

We first investigate the negative group index of refraction in graphene-based HMMs [23]. The frequency dependent effective group and phase index of refraction with angles of incidence is shown in Fig. 9.3a, b respectively. It is clear that the group index results positive in elliptical region and negative in hyperbolic region at far below the critical frequency. Note that the group index is completely zero in hyperbolic band at normal incidence. However, negative group index spectral band increases when the angle of incidence is varied from 20° to 80°. The phase index is completely positive in both hyperbolic and elliptical bands. It turns out that the graphene-based HMM provides negative group and positive phase index of refraction at oblique incidence, which allows forward wavefront propagation and negative energy refraction. The two-dimensional (2D) map of group index of refraction for chemical potential, $\mu =$ 0.2 eV and $\mu = 0.8$ eV is shown in Fig. 9.3c, d respectively. As can be seen, the negative group index spectral band increases with increasing chemical potential.

The calculated fractions of incident light transmitted through the air-HMM interface for different angles of incidence and polarizations (TM and TE) are shown in Fig. 9.4a, b. We solved Fresnel's equations to calculate the transmitted intensity. It is evident that there is a discontinuity in transmitted intensity at the critical frequency (24 THz) for both TM and TE polarizations. More importantly, different transmission minima are obtained particularly in hyperbolic region for TM polarization at oblique incidence. The frequencies at which transmission dips obtained are correspond to the maximum negative group index values of graphene-based HMM. However, it is

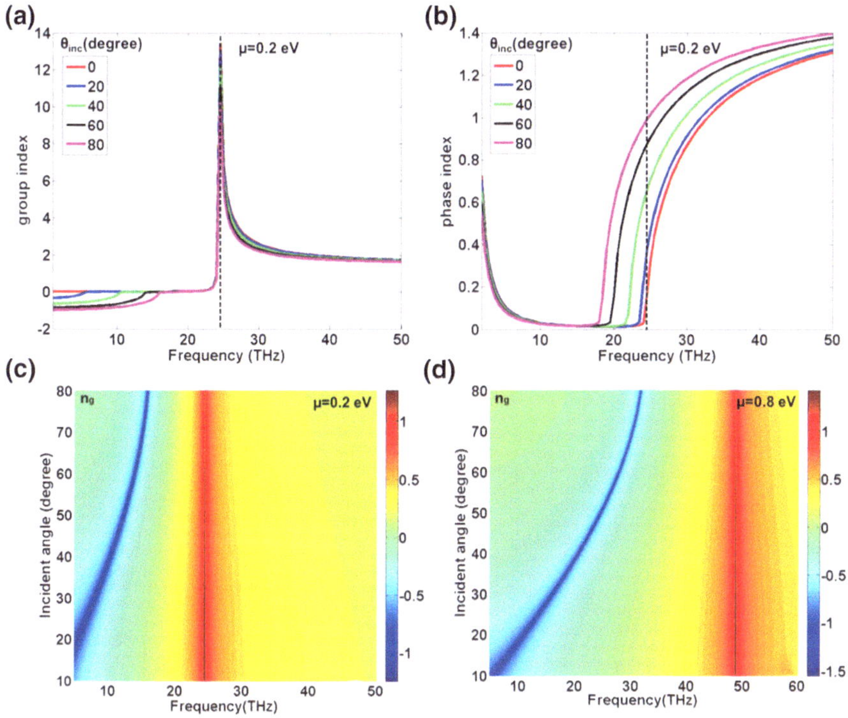

Fig. 9.3 Frequency dependent **a** group and **b** phase index of refraction for different angles of incidence. In **a** and **b**, $t_d = 50$ nm and $\mu = 0.2$ eV. 2D map (incidence angle vs. frequency) of group index of refraction for **c** $\mu = 0.2$ eV, and **d** $\mu = 0.8$ eV. Reproduced from Sreekanth et al. [23], with the permission of AIP Publishing

clear from Fig. 9.4b that such behavior is not observed in the case of TE polarization. Therefore, it can be concluded from the transmission data that the graphene-based HMM supports negative refraction for TM polarization at oblique incidence. The 2D map (incidence angle vs. frequency) of the transmitted intensities of TM and TE polarization is shown in Fig. 9.4c, d respectively.

9.4.2 Negative Refraction in Topological Insulator-Based HMMs

We then investigated negative refraction in Bi_2Se_3 in both *type I* and *type II* hyperbolic bands [24]. The Bi_2Se_3 with indefinite permittivity $(\varepsilon^x(\omega) \cdot \varepsilon^z(\omega) < 0)$ exhibits negative refraction with respect to $\boldsymbol{S}$ for TM polarization. However, both $\boldsymbol{k}$ and $\boldsymbol{S}$ refract normally in the case of TE polarization because the TE polarized light does not experience anisotropy.

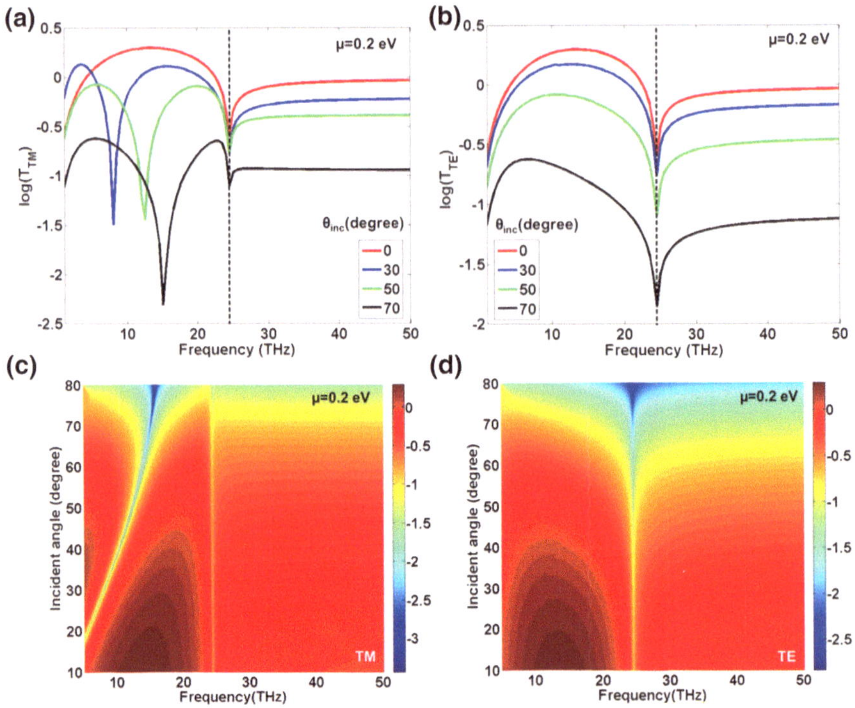

Fig. 9.4 Angle dependent transmission spectra for **a** TM-polarization, and **b** TE-polarization. 2D map (incidence angle vs. frequency) of transmission for **c** TM-polarization, and **d** TE-polarization. In **a**–**d**, $\mu = 0.2$ eV. Reproduced from Sreekanth et al. [23], with the permission of AIP Publishing

Figure 9.5a, b respectively show the calculated group and phase index of refraction with angles of incidence. In both *type I* and *type II* hyperbolic bands, the group index is negative, and the phase index is positive. It indicates that Bi_2Se_3 exhibits negative refraction with respect to $\boldsymbol{S}$ and positive refraction with respect to $\boldsymbol{k}$, allowing forward wavefront propagation and negative energy refraction. More specifically, in the *type I* hyperbolic band all angle negative group index of refraction is obtained, however, negative group index of refraction at higher incidence angles (above 35°) is obtained in *type II* hyperbolic band. In contrast to graphene-based HMM, Bi_2Se_3 shows both angle independent and angle dependent negative energy refraction due to the existence of both *type I* and *type II* hyperbolic bands.

To further emphasis the observed negative group index feature of Bi_2Se_3, the fraction of incident light reflected and transmitted at the air-Bi_2Se_3 interface as a function of incidence angle is calculated. A 2D map of the reflected and transmitted intensity as a function of incidence angle and frequency is shown in Fig. 9.5c, d respectively. Here, both the reflected and transmitted TM polarized spectra are normalized with TE polarized spectra. In particular, the solid blue-yellow curve shown in Fig. 9.5c ($\omega < 1.9$ THz and $\omega > 4.9$ THz) represents the frequency dependence of the Brewster angle of TM-polarized light on the Bi_2Se_3 crystal, which is dielectric across

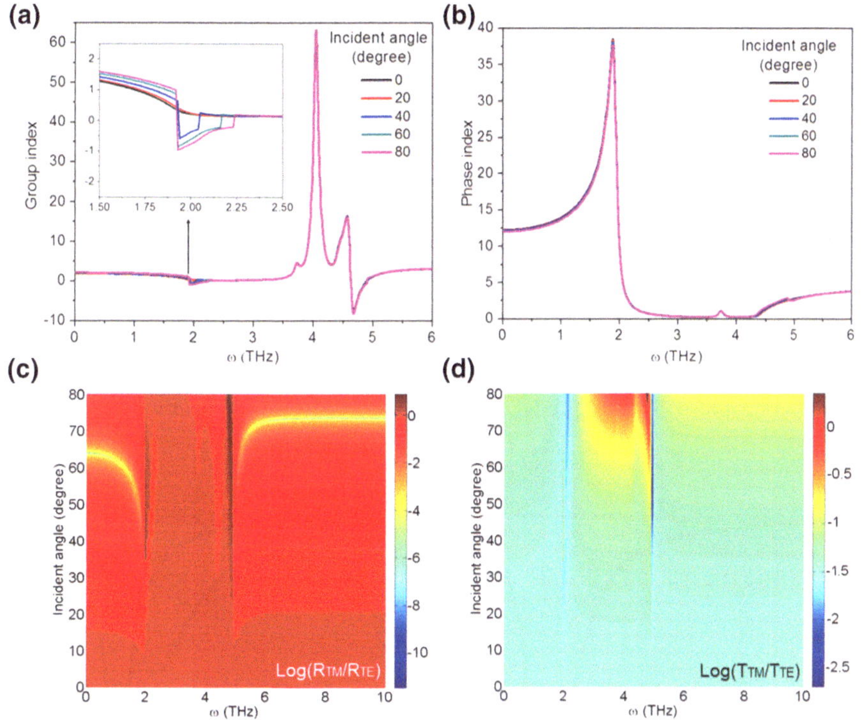

Fig. 9.5 Frequency dependent **a** group and **b** phase index of refraction for different angles of incidence. 2D map of **c** reflection and **d** transmission spectra as a function of incidence angle and frequency. Reproduced with permission from Sreekanth and Simpson [24]

this frequency band. It should be noted that an angle dependent discontinuity in the reflection (reflection peak) and transmission spectra (transmission dip) are obtained in *type I* and *type II* hyperbolic bands. This is due to the fact that Bi_2Se_3 supports negative group index refraction in both hyperbolic bands. In short, it can be concluded that the quintuple-layered Bi_2Se_3 provides extreme anisotropy and supports negative group index of refraction for TM polarization in both hyperbolic bands.

9.5 Excitation of BBP Modes of Graphene-Based HMMs

As discussed in Chap. 8, the BBP modes of the HMMs can be excited using a coupling technique. Here, we employ attenuated total reflection (ATR) method (Otto configuration) to investigate the bulk plasmon modes in the graphene-based HMM. The Otto configuration (Fig. 9.6a) consists that an HMM is placed at a distance, $d = 1\ \mu m$ from the semi-cylindrical germanium prism of refractive index, $n_p = 4$ [23]. The reason behind the selection of a semi-cylindrical high refractive index prism is that it can be used to access the wide range of effective indices and to effectively match the momentum between the incident light and the guided modes of the HMM. To calculate the angular reflectance spectra for different chemical potential and frequencies, we used the well-known Transfer matrix method.

According to Fig. 9.6, the locations of minimum reflected intensity are directly indicating the modal index ($n_p \sin(\theta_{inc})$) of the guided modes (bulk plasmon modes) excited via ATR method. As shown in Fig. 9.6a, the designed structure supports a fundamental mode with an effective index of 3.73, for $f = 10$ THz and $\mu = 0.2$ eV. Remarkably, the respective frequency appears to be in the negative group index region of the HMM. We further studied the variation of modal index with excitation frequency and chemical potential to investigate the behavior of bulk plasmon modes in the hyperbolic region. The results shown in Fig. 9.6b indicate that the reflectance spectra became wider and modal index decreases when the excitation frequency is

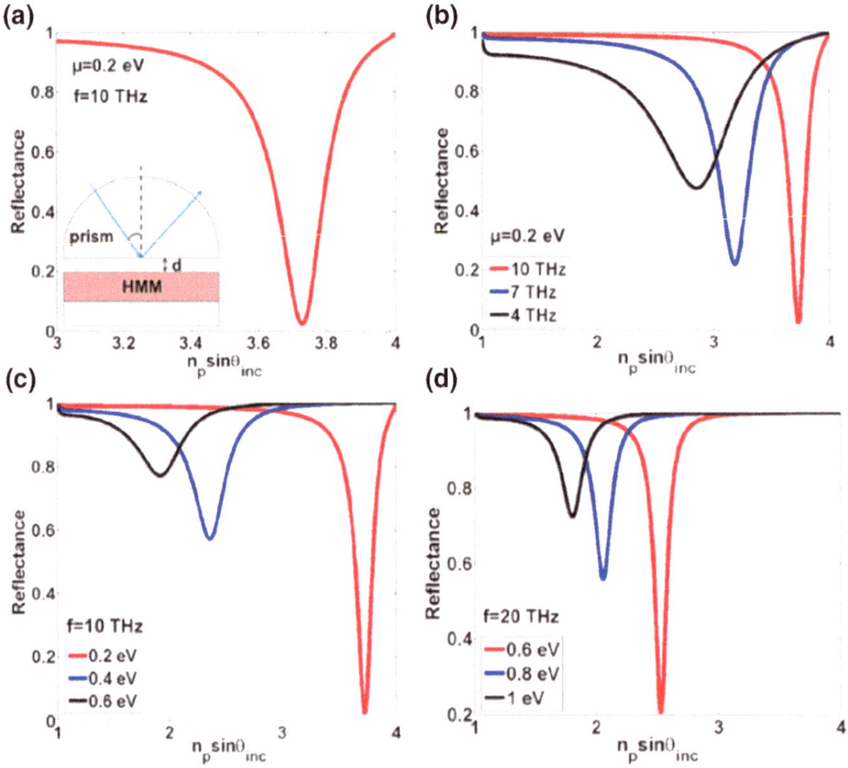

Fig. 9.6 **a** Reflectance spectrum as a function of parallel wavevector, for $f = 10$ THz and $\mu = 0.2$ eV. The inset of **a** shows the Otto configuration. **b** Angle dependent variation of modal index **b** with frequency, and **c** and **d** with chemical potential. In **b**, $\mu = 0.2$ eV. In **c** and **d**, $f = 10$ THz and $f = 10$ THz, respectively. Reproduced from Sreekanth et al. [23], with the permission of AIP Publishing

decreased from 10 to 4 THz for $\mu = 0.2$ eV. In addition, it is possible to observe the reflectance minima for higher frequencies when the chemical potential increases (Fig. 9.6c, d), which is due to the increase of negative group index spectrum of HMM with increase in chemical potential. Also note that the modal index decreases with increase in chemical potential for the same frequency. It is important to note that no such reflectance minimum is obtained when the excitation frequency is selected above from the negative group index spectral band of the hyperbolic region. The observed behavior of bulk plasmon modes in the hyperbolic band, as a function of excitation frequency and chemical potential further evidences the negative refraction features of graphene-based HMM at THz frequencies.

9.6 Active Hyperbolic Metamaterials Based on Topological Insulator and Phase Change Material

We further propose that topological insulator-dielectric insulator based superlattice structures can be used to widen the negative group index spectral band [24]. As shown in the inset of Fig. 9.7a, the proposed HMM superlattice structure comprises of alternating thin layers of Bi_2Te_3 and GeTe. The plasmonic response of topological insulator-dielectric insulator multilayer structures have been previously investigated [25]. Here, we used Bi_2Te_3 as the TI because the crystal structure of Bi_2Te_3 is similar to that of a Bi_2Se_3 whilst GeTe is a phase change material. Indeed, topological insulator-dielectric insulator superlattice structures resemble the interfacial phase change material (iPCM) structure [26]. For the calculation of the effective permittivity components of Bi_2Te_3-GeTe HMM, the permittivities of Bi_2Te_3 are obtained from the Drude-Lorentz model, $\varepsilon(\omega) = \varepsilon_\infty - \frac{\omega_D^2}{\omega^2 + i\omega\gamma_D} + \sum_{j=1} \frac{\omega_{p,j}^2}{\omega_{0,j}^2 - \omega^2 - i\omega\gamma_j}$ with $\varepsilon_\infty = 51$, $\omega_D = 207.06$ THz, $\gamma_D = 5.507$ THz, $\omega_{p,1} = 44.91$ THz, $\omega_{0,1} = 2.0095$ THz and $\gamma_1 = 0.2998$ THz. The used dielectric permittivity of GeTe is 15, which is that of an amorphous GeTe at THz frequencies [27] and the TI fill fraction is set to 0.35. In Fig. 9.7a, we show the calculated real uniaxial permittivities of Bi_2Te_3-GeTe HMM, which demonstrates broadband *type II* hyperbolic dispersion from THz to M-IR frequencies, where $\varepsilon^x(\omega) < 0$ and $\varepsilon^z(\omega) > 0$.

The calculated group index of refraction with different angles of incidence are shown in Fig. 9.7b. The negative group index of refraction is obtained for the incidence angles greater than 35° and the negative group index spectral band is widened with increasing angles of incidence. In comparison to Bi_2Se_3 HMM, a broad group index spectral band (2–5 THz) is obtained due to the broadband *type II* hyperbolic dispersion of the Bi_2Te_3-GeTe HMM. A false color plot of calculated fraction of incident light reflected at the air/Bi_2Te_3-GeTe HMM interface as a function of frequency and incidence angle is shown in Fig. 9.7c. A solid blue-yellow curve obtained above 30 THz represents the Brewster angle condition for TM-polarization since Bi_2Te_3-GeTe HMM behaves like an isotropic dielectric medium above 30 THz. It should be noted that a discontinuity in reflection spectra (reflection peak) with incidence angle (35°–80°) is obtained in the negative group index spectral band (2–5 THz). The observed spectral feature matches very well with the discontinuity in reflection spectra obtained in the *type II* hyperbolic band of Bi_2Se_3 HMM.

Another interesting feature of the Bi_2Te_3-GeTe HMM is that the hyperbolic dispersion properties can be tuned by switching the structural phase of the GeTe from amorphous to crystalline. The uniaxial permittivity data shown in the inset of Fig. 9.7d suggests that both *type I* and *type II* hyperbolic bands are present in Bi_2Te_3-GeTe HMM when the GeTe is in the crystalline phase. In the calculation, the dielectric permittivity of crystalline GeTe used at THz frequencies is 300 [27]. The frequency dependent group index spectrum is shown in Fig. 9.7d. It is clear that an angle independent positive and negative group index of refraction is obtained in *type II* and *type I* hyperbolic band, respectively. The physical reason for the obtained positive group index in the *type II* hyperbolic band is due to the fact that the uniaxial permittivity component, $\varepsilon^z(\omega)$ of crystalline Bi_2Te_3-GeTe HMM is considerably larger after the amorphous to crystalline phase transition of GeTe. These results indicate that the proposed Bi_2Te_3-GeTe HMM is an alternative reconfigurable HMM for THz

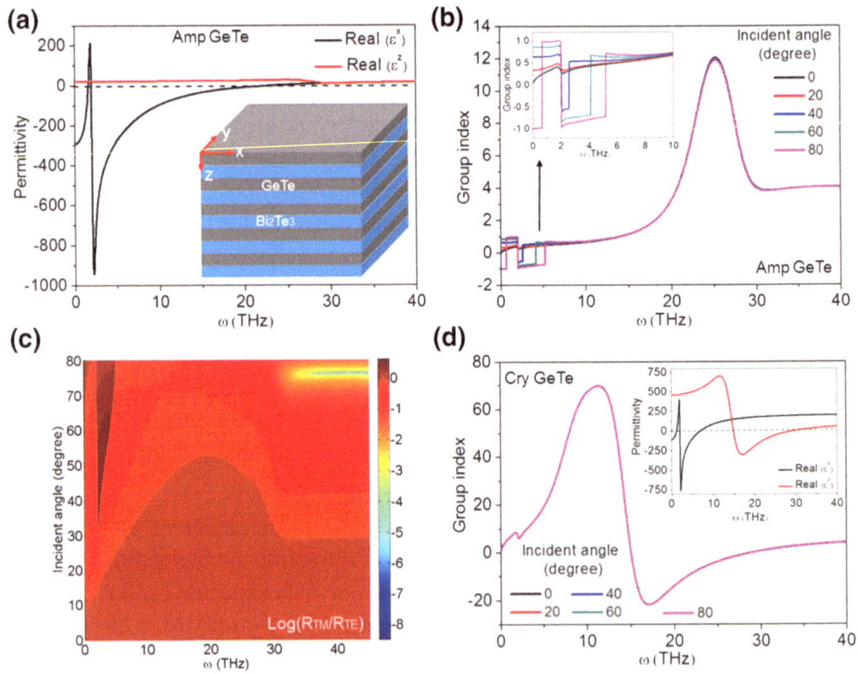

Fig. 9.7 **a** Real parts of the uniaxial permittivity components of Bi_2Te_3-GeTe HMM. Inset shows the schematic of proposed Bi_2Te_3-GeTe HMM where GeTe is in amorphous phase. **b** Variation of group index of refraction with frequency when GeTe is in amorphous phase. **c** 2D map of reflection spectra as a function of incidence angle and frequency. **d** Variation of group index of refraction with frequency when GeTe is in crystalline phase. Reproduced with permission from Sreekanth and Simpson [24]

and mid infrared frequencies since the optical properties can be tuned by switching the phase of the GeTe from amorphous to crystalline as well as by doping the surface states of Bi_2Te_3.

References

1. Sreekanth KV, Biaglow T, Strangi G (2013) Directional spontaneous emission enhancement in hyperbolic metamaterials. J Appl Phys 114:134306
2. Sreekanth KV, De Luca A, Strangi G (2013) Experimental demonstration of surface and bulk plasmon polaritons in hypergratings. Sci Rep 3:3291
3. Krishna KH, Sreekanth KV, Strangi G (2016) Dye-embedded and nanopatterned hyperbolic metamaterials for spontaneous emission rate enhancement. J Opt Soc Am B 33:1038–1043
4. Xu T, Agrawal A, Abashin M, Chau KJ, Lezec HJ (2013) All-angle negative refraction and active flat lensing of ultraviolet light. Nature 497:470
5. Naik GV, Saha B, Liu J, Saber SM, Stach EA, Irudayaraj JMK, Sands TD, Shalaev VM, Boltasseva A (2014) Epitaxial superlattices with titanium nitride as a plasmonic component for optical hyperbolic metamaterials. Proc Natl Acad Sci USA 111:7546–7551
6. Boltasseva A, Atwater HA (2011) Low-loss plasmonic metamaterials. Science 331:290–291
7. Hoffman AJ, Alekseyev L, Howard SS, Franz KJ, Wasserman D, Podolskiy VA, Narimanov EE, Sivco DL, Gmachl C (2007) Negative refraction in semiconductor metamaterials. Nat Mater 6:946
8. Chang YC, Liu CH, Liu CH, Zhang S, Marder SR, Narimanov EE, Zhong Z, Norris TB (2016) Realization of mid-infrared graphene hyperbolic metamaterials. Nat Commun 7:10568
9. Iorsh IV, Mukhin IS, Shadrivov IV, Belov PA, Kivshar YS (2013) Hyperbolic metamaterials based on multilayer graphene structures. Phys Rev B 87:075416
10. Guan S, Huang SY, Yao Y, Yang SA (2017) Tunable hyperbolic dispersion and negative refraction in natural electride materials. Phys Rev B 95:165436
11. Dai S, Ma Q, Liu MK, Fei Z, Goldflam MD, Wagner M, Watanabe K, Taniguchi T, Thiemens F, Keilmann F, Janssen GCAM, Zhu SE, Jarillo-Herrero P, Fogler MM, Basov DN (2015) Graphene on hexagonal boron nitride as a tunable hyperbolic metamaterial. Nat Nanotechnol 10:682–686
12. Krishnamoorthy HNS, Zhou Y, Ramanathan S, Narimanov E, Menon VM (2014) Tunable hyperbolic metamaterials utilizing phase change heterostructures. Appl Phys Lett 104:121101
13. Prayakarao S, Mendoza B, Devine A, Kyaw C, van Dover RB, Liberman V, Noginov MA (2016) Tunable VO_2/Au hyperbolic metamaterial. Appl Phys Lett 109:061105
14. Grigorenko AN, Polini M, Novoselov KS (2012) Graphene plasmonics. Nat Photonics 6:749
15. Jablan M, Buljan H, Soljacic M (2009) Plasmonics in graphene at infrared frequencies. Phys Rev B 80:245435
16. Falkovsky LA, Pershoguba SS (2007) Optical far-infrared properties of a graphene monolayer and multilayer. Phys Rev B 76:153410
17. Vakil A, Engheta N (2011) Transformation optics using graphene. Science 332:1291
18. Hasan MZ, Kane CL (2010) Colloquium: topological insulators. Rev Mod Phys 82:3045
19. Di Pietro P, Ortolani M, Limaj O, Di Gaspare A, Giliberti V, Giorgianni F, Brahlek M, Bansal N, Koirala N, Oh S, Calvani P, Lupi S (2013) Observation of Dirac plasmons in a topological insulator. Nat Nanotechnol 8:556–560
20. Wu JS, Basov DN, Fogler MM (2015) Topological insulators are tunable waveguides for hyperbolic polaritons. Phys Rev B 92:205430
21. Cheng W, Ren SF (2011) Phonons of single quintuple Bi_2Te_3 and Bi_2Se_3 films and bulk materials. Phys Rev B 83:094301

22. Hoffman AJ, Sridhar A, Braun PX, Alekseyev L, Howard SS, Franz KJ, Cheng L, Choa FS, Sivco DL, Podolskiy VA, Narimanov EE, Gmachl C (2009) Midinfrared semiconductor optical metamaterials. J Appl Phys 105:122411
23. Sreekanth KV, De Luca A, Strangi G (2013) Negative refraction in graphene-based hyperbolic metamaterials. Appl Phys Lett 103:023107
24. Sreekanth KV, Simpson RE (2019) Super-collimation and negative refraction in hyperbolic Van der Waals superlattices. Opt Commun 440:150–154
25. Lai YP, Lin IT, Wu KH, Liu JM (2014) Plasmonics in topological insulators. Nanomater Nanotechnol 4:13
26. Simpson RE, Fons P, Kolobov AV, Krbal M, Yagi T, Tominaga (2011) Interfacial phase-change memory. Nat Nanotechnol 6:501–505
27. Kadlec F, Kadlec C, Kuzel P (2012) Contrast in terahertz conductivity of phase-change materials. Solid State Commun 152:852–855

MIX
Papier aus verantwortungsvollen Quellen
Paper from responsible sources
FSC® C105338

If you have any concerns about our products,
you can contact us on
ProductSafety@springernature.com

In case Publisher is established outside the EU,
the EU authorized representative is:
Springer Nature Customer Service Center GmbH
Europaplatz 3, 69115 Heidelberg, Germany

Printed by Libri Plureos GmbH
in Hamburg, Germany